Bringing Back the Birdsong

Dedicated to the late Ray Richards, our literary agent, the father of this book, and a supportive friend over many decades.

First published in 2013 by New Holland Publishers (NZ) Ltd
Auckland • Sydney • London • Cape Town

www.newhollandpublishers.co.nz

218 Lake Road, Northcote, Auckland 0627, New Zealand
Unit 1, 66 Gibbes Street, Chatswood, NSW 2067, Australia
The Chandlery, Unit 114, 50 Westminster Bridge Road, London, SE1 7QY, United Kingdom
Wembley Square, First Floor, Solan Road, Gardens, Cape Town 8001, South Africa

Publishing manager: Christine Thomson
Editor: Matt Turner
Design: Kate Barraclough

Cover image: Whale Bay, west view
Thumbnail strip, from left to right:
Red moki *Cheilodactylus spectabilis*
Tui *Prosthemadera novaeseelandiae*
Yellow admiral butterfly *Vanessa itea*
Pohutukawa bloom *Metrosideros excelsa*
Back cover, top: Rata *Metrosideros robusta* and kanuka *Kunzea ericoides*
This page: Splendid perch male *Callanthias australis*
Contents page: Fernbird matata *Megalurus punctatus*

National Library of New Zealand Cataloguing-in-Publication Data

Doak, Wade
Bringing back the birdsong : two dedicated conservationists work to restore the natural balance to their slice of coastal New Zealand / Wade and Jan Doak.
Includes bibliographical references and index.
ISBN 978-1-86966-401-5
1. Conservation of natural resources—New Zealand—Whangarei.
2. Nature—Effect of human beings on—New Zealand—Whangarei.
I. Doak, Jan. II. Title.
333. 91716099316—dc 23

10 9 8 7 6 5 4 3 2 1

Colour reproduction by SC (Sang Choy) International Pte Ltd, Singapore
Printed in China by Toppan Leefung Printing Ltd, on paper sourced from sustainable forests.

Bringing Back the Birdsong

Two dedicated conservationists work to restore the natural balance to their slice of coastal New Zealand

WADE AND JAN DOAK

NEW HOLLAND

Wade tests an early Mako Shark amphibious box camera in Christchurch's Olympic swimming pool, 1958.
JIM WHITE

INTRODUCTION

With a half century of undersea exploration under our weight belts Jan and I came ashore. We felt driven to assemble the *big picture*; a slice of life on our planet; a local portrait of New Zealand wilderness above and below.

Until a certain eureka experience brought me ashore I was totally preoccupied from the age of 14 with exploring under the sea. My generation was the first to have the privilege of escaping gravity and first-finning on a blue planet very new to humans. The invention of the aqualung gave us the key to an unknown world.

I recall a scuba dive with Kelly Tarlton in Lyttelton Harbour at a location that has now been reclaimed: Golins Bay. Visibility is so poor we have only centimetres between our masks and the encrusting rock walls. But we are enchanted. This *is* another world. We do not recognise any of that patchwork quilt of life forms at our fingertips. *I think that is what induced us to become naturalists.* The unknown, and having to look very closely; and a heightened feeling of privilege at entering such a special place.

In 1959, while attending Canterbury University, I established, with my diving friends, a bi-monthly diving journal. For the next 15 years I edited and published *Dive* magazine, aided by Jan and other diving mates, Jaan Voot, Keith Gordon and Kelly Tarlton. Its pages document every aspect of the developing New Zealand diving world I could scrounge together. The pages of my diving magazine tell of our many adventures: south to Stewart Island; west into the Southern Alps for New Zealand's first ice dive in a frozen lake; and a whole series of expeditions to parts of the tropics little known to divers.

The bombshell was a letter from our diving buddy Kelly Tarlton, sent from Whangarei in the far north. His Telecom job had transferred him up there. Kelly insisted we come and dive the Poor Knights Islands, beyond question New Zealand's finest undersea province. Located 22 km off the coast from Tutukaka Harbour, they are bathed by a warm, nutrient-laden current from the north. This would be the undersea world of our dreams – and beyond, right up to the moment of writing these words. In 1963 Jan and I

married and shifted north to be near paradise. We ended up living in a cottage 3 m from the mangrove-clad banks of the Matapouri River on the Tutukaka Coast. (After a flood nearly took our home to sea, in 1978, we shifted to a bush-clad hilltop 60 m above the Ngunguru River; 18 hectares of wilderness.)

So for many years I devoted my time obsessively to documenting life below the tideline. Greatly assisted by my diving wife, I ventured widely and wrote a bookshelf of titles about shipwrecks, marine life, the behaviour of fishes, encounters with dolphins and whales, exploring coral reef communities: if it was wet I was rapt. The topside world, while much appreciated, was mainly a place to recover warmth before heading below again.

This is the sort of picture I was writing about marine life when my world tipped over:

> *Waving from a crevice in the richly encrusted sea cliff are twin pairs of long white antennae. Within I find a banded coral shrimp. Close by is its partner. Coral shrimp couples live in the same shelter for several years. Their boldly marked bodies, striped like a barber's pole in red and white, and their showy antennae have a special purpose.*
>
> *Most shrimps and crabs are very secretive during the day because fishes regard them as favourite items of diet. But no fish would ever eat a coral shrimp. These are the surgeons of the reef. Their crevices are cleaning stations. (Barbers too once acted as minor surgeons.) Attracted by the prominent feelers and bold markings, reef fish congregate at these clinics and await treatment. With specially adapted pincers, very like a surgeon's forceps, the nimble shrimps remove sea lice and fungal growths from the gills, fin bases and skin of fishes. There is immense trust all around. The shrimps even enter the mouths of big predators such as groupers or moray eels, to clean their teeth. If I open my mouth a coral shrimp may even enter to clean mine!*

ABOVE: First ice dive in New Zealand at Lake Ida; Wade on brink, 1958.

OPPOSITE: Banded coral shrimp *Stenopus hispidus* and greenhood orchid *Pterostylis banksii*, patterns that connect.

One day in early spring of 1974, when it was too stormy to dive, I set out for a stroll in the rainforest near our home in Matapouri Bay. My ecologist diving companion Dr Tony Ayling, who had a passion for botany and marine life, alerted me to some tiny life forms on the forest floor. For the first time I saw the fragile greenhood orchids that I'd been treading on all my life. As I knelt close to them a thought stirred in my mind. A memory from beneath the sea. Those long, pale pink tendrils reminded me of the coral shrimp's feelers. But then another realisation struck me: the elongate sepals would serve to attract insects to land on the green platform beneath them and pollinate the orchid. Similar pattern: similar purpose.

From that moment wherever I rambled in the rainforest patterns emerged with which I was already familiar after years of exploring the sea. I found that by comparing patterns from two very different worlds, ocean and rainforest, I could arrive at a basic blueprint, the pattern common to both. With a bagful of cameras and lenses I began to study and compare life patterns in the mangroves around our river edge cottage and in the nearby headland forests with those I knew so well in the sea, coming to appreciate them as ecosystems just as fascinating and diverse as the reefs I had been exploring in the tropics and those of the Poor Knights.

In the forest I used the same camera techniques and lighting as when diving so that I could focus and pry into all the wettest, darkest places. As I pressed the trigger I even found myself holding my breath to avoid the noise of bubbles! I was *land diving*. I came to see that above and below the basic patterns of life are mirrored. Whether I climbed from the sea's edge up cliffs into the headland forest or descended a steep undersea wall, in both ecosystems I encountered a great diversity of life forms clinging to a firm substrate, struggling for standing room, branching upwards or outwards in their quest for energy.

When Jan and I emerged from the sea and entered all the lovely forest worlds of Northland, we faced a massive challenge. As amphibians we

Wade with movie camera on tripod, Becks Bay: ritual of the orchid faith.

had only a smattering of knowledge about what is truly native to these shores. But our passion grew to an obsession as we strove to match our terrestrial understanding with our knowledge of the world beneath the waves. We explored our neighbourhood in its entirety, all habitats, both above and below, from the Bougainville Promontory adjoining Whananaki in the north, to Goat Island/Harakeke in the south and as much of the hinterland forest-clad hills and valleys as we could gain entry to. From detailed scrutiny of the lovely Whale Bay Reserve and Matapouri River our focus shifted over the hills to the much more challenging complexities of the Ngunguru River catchment. We were discovering a slice of the planet: our own local slice of wild New Zealand. Now, as septuagenarians, we have developed an archive of its flora and fauna and recorded many of its stories.

A new era began around 2007 when we embarked on an intensive regime of pest control. For many years we had battled to have our beloved Poor Knights Islands recognised as a sanctuary. In later years, Jan and I escorted a TV team through that magnificent marine reserve and felt intense joy as they marvelled at the swarming marine life around us. I told them that this ocean community was

much older than the islands themselves: we were time travelling …

Now, in the protected forest surrounding our Ngunguru home, we are seeing the ancient world of Gondwana reasserting itself. Like a photograph developing in a darkroom tray a big picture is emerging: the real Aotearoa/New Zealand, above and below, a magical world that our grandchildren are exploring. *A world we dream could extend to our entire country.*

So this is the story of how Jan and I set out to create a record of the wilderness in our locality: from undersea reef to mangrove world; across the intertidal, up the sea cliff and into headland forest. When we ventured further abroad we learnt to our dismay that New Zealand forests are in trouble. So few birds. So many endangered plants. Our own pest-protected land is much richer. Visits to sanctuaries staggered us at the fecund essence of a pest-free New Zealand. So we resolved to restore our land, establishing rare and endangered plants and joining forces with neighbours to control harmful aliens.

We never knew what we were in for. A major part of our book is a nationwide story of the amazing explosion of wildlife that has occurred, expanding as I write. But first we had to learn what plants and insects, birds and trees actually belong to the ancient community we are so fortunate to inhabit, and how their amazing stories are so intricately connected.

Sentinel hollow puriri tree Whale Bay track. In the tree is Wade's grandson, Mark Miller.

Wade in a rare black coral tree *Lillipathes lilliei*, 60 m below the surface.
KELLY TARLTON

CHAPTER 1:

BECOMING A NATURALIST UNDER THE SEA

If you were visiting another planet swarming with unknown and bizarre life forms, becoming a naturalist would be de rigueur. And if your spaceship library had little to help you, perhaps you would start writing books about alien life. When I reach back some 50 years I can see why early scuba explorers had to become naturalists – as are so many divers today. As space rockets gave access to a sterile moon, Jacques Cousteau's aqualung or scuba technology enabled ordinary people to explore a new world teeming with alien creatures.

Until the 1960s the vibrant sea life of steep-walled ocean islands was virtually unknown. Helmet divers could not operate in such conditions and sought only to salvage or construct. Scientists had intensively studied the intertidal to the low-tide mark. Beyond that they were blind and had to rely on remote techniques such as lowered dredges, which were very limited on 'foul ground'.

In the early 1960s, there was an avalanche of discoveries as Kelly Tarlton, Jan Doak and I, along with a small group of scuba divers, first-finned the Poor Knights Islands, an undersea realm humans had never seen. Looking back I can see that as pioneers we were profoundly affected. The impact of visiting unknown vistas and wildlife on the other side of the tideline leads to clichés that were very real to us at that time. In Cousteau's words, we were all oceanauts exploring a blue continent.

Our initial task was to identify and label: the taxonomic phase. In those days diving scientists were very rare. We had to find ways to bring the sea to the laboratory. We brought up samples, preserved them in formalin or alcohol and sent them off to institutions. But there was a need to see the bigger picture: the context or habitat in which these plants and animals lived. To achieve this we had to do a survey: to photograph a strip of sea cliff, metre by metre, down to the tranquil sea floor: the wall of mouths and all its transitions as light levels and wave action diminish.

THE GIANT STAIRCASE

In December 1965 Kelly and I, along with a band of divers, landed on the Poor Knights Islands, a land sanctuary 22 km out from my home on the Northland coast. We had special permission to carry out a trial submarine survey there. Our project entailed the construction and fastening of a numbered grid to a cliff face. The grid looked like a huge ladder 1.8 m wide and 1.5 m between 'rungs'. Once photographed it recorded a 40 m strip of sea cliff, with numbered quadrats enabling identification of any specimen or area of interest to biologists, and showing the transitional process with increasing depth.

There is only one point of access on the steep-cliffed northern Knight Tawhitirahi: our campsite and diving area could only be Landing Bay, fortunately an excellently sheltered diving location. Directly beneath it the cliffs plunge to 45 m. Visibility is often around 30 m, rarely less than 18 m.

Despite the 30 tanks we landed, and a second batch brought out when weather permitted, we had to ration ourselves to two scuba dives each per day.

Tawhitirahi, northern Poor Knights Island, east view.

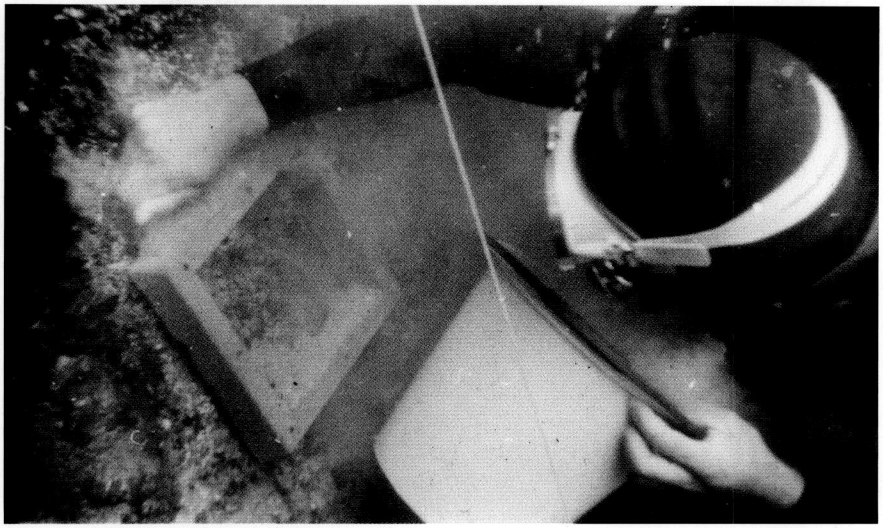

This meant we had to devote most of our time to the project. While selecting a site, Kelly noticed a peculiar shaped object on the sandy sea floor 30 m out from the cliff; but it would have to wait until we completed the grid.

A 90-degree cliff face was selected. A series of dives soon had the grid unravelled and fastened in place, and a concrete plaque set into the cliff bearing details of the experiment for future identification and comparison.

Using his massive electronic flash rig, Kelly descended the grid, while I held the number steady for his camera at each quadrat. The most difficult shots were those near the surface as the moderate swell, rebounding off the cliffs, tossed the kelp fronds into a frenzy, jostling both photographer and assistant. As we dropped down into the clear still depths that strange paradox of diving became apparent: the surface is a hostile, alien world and the quiet depths produce a feeling of oneness and tranquillity. This is where a diver belongs.

Having photographed the quadrats down to the last one at 42 m, just 3 m above the sandy bottom, we sneaked enough of our precious air supply to take a look at the object on the sand floor behind us. A wreck! As we swam over the bow, past the diesel engine and its sponge colony to the gently tilting stern, it was soon apparent that this was a burnt-out hull of a fishing vessel. (A radio call to land confirmed it as an eight-year-old 'loss' by fire, well covered by insurance.) While not spectacular as wrecks go, the marine growth and fish life that had adopted it made wonderful camera subjects.

Our next step was to cement concrete frames for a future experiment to the cliff face at intervals on the grid. The rock face beneath was scraped clean and fibreglass netting fixed over the frames, so that the minute sea life beneath could be studied, ungrazed by sea urchins, molluscs or fishes.

OPPOSITE: Wade hovers over a deep wreck in Landing Bay, Poor Knights Islands.
KELLY TARLTON

Jaan Voot lays cement on the Giant Staircase grid.
KELLY TARLTON

To make cement on this remote, beachless island we had to bring up buckets of sand from the sea floor 45 m down, wash it in fresh water and dry it before mixing. By trial and error we soon found a method of transferring the cement from bucket to rock face without losing it all. The biggest problem, Jaan Voot said, was to fight off the triggerfish that were gulping it down at every opportunity!

FIRST POOR KNIGHTS NIGHT DIVE

Having completed the project we were able to use our remaining air for exploring. To date, no one had dived here at night, far out in the South Pacific on the edge of the great Kermadec Trench abyss. On day trips we had already discovered many hitherto unknown species of fish. The Auckland Museum was thrilled with all the new mollusc species we had brought up. However, we knew we had only scratched the surface. There were still many fish we couldn't get near, and shellfish are much more active at night. In readiness, we had brought a variety of underwater lights for deepwater use.

We had all night-dived often but never to such depths. As the four of us slipped into the black waters and grouped ready for descent, Byron Anderson sat on the cliff ledge, holding a powerful light to give us a reference point. There was no moon. We had tired of waiting for ideal conditions and at least it was reasonably calm, except for the back-surge from the cliffs.

Dropping 45 m down a cliff face in a slow fall while pencilling the blackness with our torch beams was like taking an eerie elevator ride. We

OPPOSITE: A sharp-nosed pufferfish *Canthigaster callisterna* rests in a sponge at night.

TOP: A community of reef fish: Sandagers wrasses *Coris sandeyeri* — gaudy male, white females, banded juveniles; combfish *Coris picta;* and goatfish *Upeneichthys porosus*.

BOTTOM: Pacific deer cowrie *Cypraea vitellus* active at night.

knew these rocks and caves so well by day, but now they were surrealistic. In the utter darkness the small world of light created by our torches drew us close together and even closer to the rocks and crevices. Piled on each other, drowsy and docile, pink maomao and moray eels stirred their fins sleepily in our torch beams. The maomao swims continually during the day, never lounging on rocks. Now its bright, lolly-pink skin was white blotched, adding to its appearance of inertness.

On the sandy bottom at 36 m were goatfish, scuffing the sand gently with their barbels. Raising minor sandstorms, huge hermit crabs scurried through the dunes like camels. Our flashbulbs flared in the dark void like lightning.

We realised that fish behaviour could only be studied over a 24-hour period. The species we were swimming among were obviously living on a shift basis: those that swam by day were now lodged in the same rock clefts that are daytime dormitories for the night feeders.

Over a piece of coral slid a gleaming chestnut-brown cowrie shell. Jeff Pearch and I closed on it like a gemstone. This was the first live specimen recorded in New Zealand and only the third ever found here! In the course of this one brief deep night dive we had learnt more than in a score of daylight dives.

After seven years of Poor Knights exploration, my marine invertebrate book *Beneath New Zealand Seas* (1972) could draw on research by the first generation of young diving scientists, my companions. I was able to write a formal ecological description of this sea-cliff descent. In our ancient ocean world, this should remain basically meaningful anywhere on the globe. The name we gave the location – Giant Staircase – is still in use.

POOR KNIGHTS ECOLOGY

In the next phase of Poor Knights exploration, with our own high-speed dive boat, Jan and I were able to attract more and more scientists to join our explorations: a new breed of young PhD students, all fully trained scuba divers, who became our companions and lifelong friends. While they undertook studies of marine invertebrates and fishes, I documented their work with word and camera for future books.

In late June in 1970 Kelly and I dived at the Pinnacles, near the Poor Knights. A large grotto, which we called The Slot, harboured up to 30 hapuku, ranging in size from 10 to 60 kg. The biggest would have measured 1.8 metres. I desperately

OPPOSITE: Hapuku *Polyprion oxygeneios*, a wide-ranging, gregarious predator, now an extremely rare sight.

needed a dramatic hapuku photo for the cover of my first book on New Zealand fishes.

At 60 m I hold my breath. A very large hapuku is hurtling towards my camera. Closer and closer come the great pointed jaws, the jagged dorsal spines and the saucer-shaped eyes. I preset the focus at two metres, adjust the flashgun angle and watch through the lens. For a split second it is in focus. As I fire it veers aside, offering a more dynamic angle. Within The Slot the rest of the herd huddles, a seething mass of 'chumbling' jaws and silhouetted dorsals. At that depth we can spend only a few minutes before heading up. I have the cover shot for the first of my four New Zealand fish books: *Fishes of the New Zealand Region* (1972).

From overfishing, hapuku have now become virtually extinct within the New Zealand diving zone. Few divers will ever have the chance to see hapuku, yet they used to be common in shallow water. This is another great tragedy of my diving life, along with vanished squadrons of colossal kingfish, massive trevally schools and giant packhorse crays.

Then there was the disappearance of all the huge black coral trees Kelly and I had seen at the Pinnacles. A 6-m high tree we discovered there was probably the world's biggest – until, tragically, it vanished. Very likely the villain was a craypot line that became fouled in its branches, having to be torn loose with the power winch. This was the motivation for the huge effort put in by me and other divers to get a marine reserve gazetted around our beloved Poor Knights Islands in 1981, second in New Zealand. The decline of pink maomao and golden snapper schools appears to have been arrested. With full protection since 1998 the islands have now been transformed. I now realise, after seeing 50 years of decline in New Zealand marine life, old divers usually become keen conservationists.

Over the ensuing years I had published four books on fish identification and ecology; a parallel pair on marine invertebrates; four on cetaceans: whales and dolphins; and two about marine life in the tropics. I published 75 issues of my own *Dive South Pacific* magazine. Then came that life-changing eureka episode with the greenhood orchid (see page 9), triggering migration up the coastal slope and into the forest.

TOP: Packhorse lobster *Sagmariasus verreauxi*, huge packhorse crays like this are vanishing from our seas.

BOTTOM: Pink maomao *Caprodon longimanus* asleep, skin white blotched for camouflage.

Ngunguru River mouth and sandbar.

CHAPTER 2:

OUR SLICE OF WILD NEW ZEALAND

'Long stretches of sandy beaches alternate with intimate bays, steep cliffs, and deeply cut natural harbours.' So wrote *National Geographic Traveler* magazine when they rated the Tutukaka Coast second equal best coastal travel destination in the world. And, therefore, *best in New Zealand!* Embracing the Poor Knights Islands offshore they said: 'This is mainly a divers' and bush walkers' destination, and should stay that way.'

Eastward aerial view of the Ngunguru River.

One fine day in mid-July of 2011 Jan and I were invited on a coastal flight by our friends Adrian and Judy Gilbert. We especially wanted to get aerial video of this coastal community, having studied its past intensively from as far back as we can find written record, searched on foot for every relict we can locate and immersed ourselves in its natural history.

With the dramatic low lighting of winter to present the landscape at its best, everything was in our favour. The tide was brimming in all the turquoise channels of Whangarei Harbour and in minutes we were circling over our forest-whelmed home on the north bank of Ngunguru River. Approaching the river entrance from the north, Adrian enabled our lenses to drink in the powerful statement of Ngunguru Sandspit/Pimanu peninsula and the blue elbow where the estuary sweeps inland. Then we flew westward along the forested, highly indented shores, with their low-density scattering of hilltop homes, each surrounded with an emerald green sculpture made by ride-on mowers. Some reminded me of crop circles. Above Scow Landing, Jan framed up the entire tidal part of the Ngunguru River meandering out from Kiripaka to the coast. Then we flew north over the Glenbervie high country following the skeins of a freshwater stream coursing through rocky ravines with white patches of rapids here and there, until the river became a radiating series of small creeks draining the rolling Kaiatea farmland. It is a paradox that this eastward-flowing river begins its life as a westward stream.

The Tutukaka Coast, about 50 km from the city of Whangarei, lies at latitude 35° 32' South on the east coast of Northland. Roadmakers have never been able to overcome the steep terrain that defines its northern and southern limits. Whether I stand on the northern promontory tip known as the Bougainville Monument, adjoining Whananaki; or peek south through a window in forest crowning the summit of Whakairiora Mountain towards the turrets of Pataua and Whangarei Heads, I understand how geology has defined this coast and, in so doing, bestowed a blessing on all who have lived here.

A STUDY IN SEA LEVELS

When I look closely at the topography – all the cliffs, caves and rock platforms – the question arises: where did this splintery grey rock come from? The strata are so confused, twisted and bent, even juxtaposed at right angles. A cataclysm must have occurred.

Some 200 million years ago when ichthyosaurs like dolphin/crocodile hybrids prowled the oceans and dinosaurs roamed the great southern supercontinent of Gondwana, before any birds sang or any flowers bloomed on our planet, there was no New Zealand land mass – just ocean deeps. For 100 million years sediments from what is now Australia and Antarctica travelled eastwards on currents and rained down onto an abyssal plain. Eventually such a vast thickness of sand, silt, clay and gravel accumulated down there that the Earth's crust buckled. Then it heaved back up, so that deeply buried sediments, compressed and heated into varying forms of rock, were thrust above the sea. Zealandia, our ancestral continental mass, was born.

For millions more years Zealandia was eroded by rivers and streams until the ocean reinvaded two-thirds of the land mass, leaving only a series of small island groups.

Then some 20 million years ago, not long after the first whales appeared, the Pacific tectonic plate began sliding beneath the Australian plate in our vicinity. A new phase of mountain building began that continues today. Under immense pressure sedimentary rocks were crushed, sheared, sharply folded, jointed, dissected by fault lines and even turned completely upside down.

After the most recent ice age, when humans were already abroad in Australia, the polar ice caps melted and the ocean began to rise (some 130 metres vertical, culminating at our present sea level about 7500 years ago), drowning the existing valleys and creating headlands and islands from the land's outer ridges and hills. Since then the onslaught of waves has tended to straighten the coastline, cutting back cliffs and infilling bays with sand or shingle.

So a headland coast from North Cape to East Cape was created. Much of it

Middle Gable, Tutukaka Coast: history in the stratified rock.

is greywacke: a fine-grained sedimentary rock formed from gravel compacted by intense pressure and heat. The stratified rock is left with a criss-cross pattern of jointing that underlies the development of all the underwater canyons, trenches and overhangs, a pattern that extends up the cliffs, as lines of weakness along which the waves erode. On a small scale a rock surface so cracked and figured provides myriad microhabitats: crannies, holes and pools. On a grand scale it is readily moulded into a series of coves and beaches set off by frequent forested headlands and fringed with offshore reefs, stacks and small wooded islands. Wherever a river outflowed, sandy bays, mangrove estuaries, raupo swamp, sandspits and lagoons were formed.

All these processes led to a youthful coastline of immense diversity and offering a very wide range of biological habitats. Because the prevailing winds blow from the land this is a warm and sheltered coast. Certainly it is assaulted by about 20 easterly storms each year, but the headlands, adjacent islands and reefs, along with unusually deep inshore waters, modify the energy of the winds and waves and inshore currents to provide a huge variety of shoreline and subtidal conditions within short sections of coast. At some locations in less than five km from north to south, there is 40 km of shoreline to explore. All these factors combine to form the essential beauty of a headland coast.

NGUNGURU'S WEALTH OF LIFE

Like the moa, the giant eagle and the huia, which are no more, New Zealand landscapes can vanish at the hand of man too. In the words of my dear late friend Geoff Park, eminent ecohistorian: 'New Zealand's most low-lying country, estuaries and tidal inlets, has become its most altered.' For this reason I hold that the 25 km Ngunguru tidal river is now one of our country's most significant. This is not the classic river: a watercourse with valley ridges either side. Ngunguru River drains a herringbone series of forested valleys with streams that intersect at right angles all along its length. Powerful, steep promontories deflect its course in sweeping bends. There are deep, mysterious places where big fish lurk. Broad leads became long lakes when the tide is full. Navigable and once abuzz with commerce, nowadays it is a tranquil waterway. The tortuous nature of its catchment has inhibited development more than elsewhere in Northland, leaving us with rare remnants of riparian subtropical rainforest fringed with very large mangroves. Most riverbanks have been cleared and modified, *because we could*.

A salient feature of this river, an icon of its former glory, is the forest promontory where my wife Jan and I have lived since 1978 (after shifting from our Matapouri cottage): Riverlands, our neighbours call it. From our research it has the highest possible biodiversity in this district. On its southern summit, above the river, rises the only stand of large kauri trees left on any river in the

Up the headland cliff: metamorphic rock.

world. With easy water access riparian kauri were the first to fall to the axe. One day in 2008 Jan and I bit the bullet and lowered ourselves down a very steep slope through those kauri, to land on a small apron of bush screened from the river by dense mangroves. Down there, in emerald light wells, we found a thicket of rare mairehau shrubs with their exquisitely fragrant leaves. Nearby was an ancient square gin bottle of hand-blown glass, perhaps tossed there by those sailors from England who camped in this area some 170 years ago. Not far away we found the crowns of felled kauri trees, lying on mangrove mud, possibly relicts from the spar-collecting days of the HMS *Buffalo* in 1837. Since then our rewards for tackling the hardest locations on this river have often been finding rare plants that have escaped the ravages of fire and cattle. We regard such places as refugia: historic sanctuaries of the district's full biodiversity.

And dotted around our land or along the river edges we discovered vestiges of a past world, perhaps eight centuries of occupation: kumara pits, terraced dwelling sites, a canoe haul-up camber, old paths, middens of seashells, a sliver of jet-black obsidian glass, stone-age tools, pa sites …

As we had already studied the local coastline, our desire grew to understand the history of this river. It was the magnet that drew settlers to this difficult Tutukaka section of a headland coast for kauri timber, coal and pipe clay. What did its shores look like when clad in virgin forest? The earliest European botanical expert to record the luxuriance of its native forest was missionary and botanist William Colenso who, in 1839, proselytised up the Tutukaka Coast on foot. At Whareora, now on the Pataua North road, he had experienced several days of torrential rain and sleepless, insect-plagued nights.

After a most wretched night from mosquitoes, which were innumerable, covering and darkening the sides of my tent and which nothing, not even fire, smoke, nor

Kauri-clad headland in Ngunguru River: probably the last on Earth.

water, would drive away, I arose feverish and unrefreshed ... Started 11 am for Ngunguru ... Found the brooks and rivers much swollen; managed to cross by stripping and sitting on the natives' shoulders ... After walking and wading for about twelve miles, we reached that part of the river where it is navigable for canoes from the sea. [Kiripaka?] Here I found a small party of natives ... We got into our canoe, and went for some time merrily down the river, gliding swiftly along by the great flood. The hills rose precipitous and high on either side, clothed with forests to their summits, and the scenery, in many places, was most enchanting. Hitherto the axe of the feller had not come up against these eternal woods; here stood the pines in nature's lovely order, some, hoary with age, elegantly festooned and draperied with mosses and lichens and climbing ferns, as if done with fairy fingers. Some, green, brilliantly green, and pristine vigour; others dark, sombre, of every shade, all beautiful!

19th December. I arose ... breakfasted and left this place, Te Peke accompanied me. An hour's walk brought us to Tutukaka, a rather romantic little harbour. Here it was that H.M.S. Buffalo lay and took in her cargo of spars in 1838. Here also were five white men residing, who seemed by no means glad to see me ... Our route, for road there was none, lay over high, steep and densely wooded hills. The recent very heavy rains and the continual rooting of the innumerable wild pigs that infest these forests had made the loamy soil little more than deep mud, so that in some parts it was with great difficulty that we made any progress ... At 2 pm we arrived at Mata Pouri, another small harbour, with a very narrow entrance. Here travelling, even moving, was almost out of the question from the luxuriance of the untrodden jungle and the quantity of mud and water. However we persevered, Te Peke continually assiduous in assisting me in every difficult place. At 5 pm we brought up at Otakahia, a sandy beach: all hands excessively fatigued. [Sandy Bay?] I pitched my tent on the beach, scarce five yards from the ocean which was rolling majestically at my feet. I saw today for the first time that very splendid pine Dacrydium plumosum [kawakawa].

Colenso diary, 18 December 1839 at Whareora

As part of the generation who has pioneered our undersea world for over half a century, Jan and I became equally impassioned by the topside of that Tutukaka Coast. We grasped only the wet half of the picture. For amphibians the tideline is an arbitrary division. We set out to gain a complete overview of all the adjoining forests and rivers and, for context, extended our interest to all those of the north.

Sunlit Tutukaka Coast headlands: Rocky Bay with the Poor Knights Islands on the horizon.

CHAPTER 3:

A DAY ON A HEADLAND COAST

The unfolding of our investigative journey begins at dawn one day at the very threshold of spring in 1972, on an ancient beach that was dry land only 20,000 years ago. I wanted to ascend the submarine headland slope, level by level of ocean life, to emerge at the interface where land and sea collide. My camera would explore the transition of life from sea world to forest glade: starting out in the depths, moving upslope into shallower waters, and finally emerging into the tidal zone of the headland.

From our tiny cottage on the bank of Matapouri River, with a pack on my back full of camera gear, I tread a moonlit path to the cliff-edged promontory. Cramped and chill I await the sun, zoom lens at the ready. I listen to the stereo chuckle of the waves and the occasional bird cry from the bush behind me.

MOTHER OCEAN

Awaiting sunrise, I have time to reflect on the role of the ocean in our lives. Offshore a mighty abyssal river is rushing past our coast from the Deep South: the Antarctic Bottom Current, Earth's most important ocean current. The Antarctic is the pulsing heart of our planet. As cold surface water sinks to the ocean floor in the Weddell Sea it drives abyssal currents that circle the globe and have a major influence on the world's weather. How vital the sea is to all life. Traces of iodine manufactured by small red seaweeds that live on blades of brown kelp regulate the growth of all vertebrate animals. Rising as methyl iodide gas from the sea it blows ashore to enter our food chain. Without this catalyst we may develop goitre. I consider how the sea's plant plankton pastures furnish most of our oxygen, absorb most of the carbon dioxide and create a compound that forms the nucleus of raindrops: dimethyl sulphide. And via animal plankton, seabirds and fertiliser, the ocean is the source of the phosphorus in our bones.

Yellow starfish *Knightaster bakeri*.

OPPOSITE: Deep-water coastal life: Poor Knights wandering anemone, a rare undescribed species.

TOP TO BOTTOM: Soft coral *Alcyonium aurantiacum* and flask sponges *Leucetusa lancifer;* crater sponge *Geodia regina.*

OPPOSITE, CLOCKWISE FROM TOP LEFT: Clown nudibranchs *Glossodoris amoena* mating; athecate hydroid *Ectopleura* sp. and feathery hydroids *Solanderia ericopsis* hydroid sea fan; up the undersea slope to shallower water, brown kelp *Ecklonia radiata* holdfast; thecate hydroid *Solanderia ericopsis.*

DEEP COAST

Out from this fingertip promontory I scuba dive 10 storeys to an ice-age beach. Down there my camera finds an ancient community of colonial sea creatures crowding a patch reef like high-rises on Manhattan Island. Down through a cold grey void I glide . . . darker and darker it grows until it seems the sun has set and I am heading into a twilight world. Only the shotline guides me for what seems an eternity of nothingness. Then from the gloom, like the onset of a dream, 9 m below me a wonderland of strange sculptures and drowned colour emerges. I swoop down beside a plateau of rock so richly adorned it takes my breath away. I have only minutes of visiting time. My camera flash illumines the scenery.

The 20-cm tall column and flowerlike disc of a wandering anemone; nearby an elegant urn-shaped sponge; golden candelabra sponges, pink gorgonian fans with small beaded polyps; delicate water-inflated orange soft corals; cup corals; a slender armed, canary yellow starfish; sponges shaped like porcelain vases, goblets, lattice-work trellises and scarlet deer antlers.

On this plateau everything has a perfect symmetry such as I have never seen on the steep cliffs of offshore islands like the adjacent Poor Knights. Nourished by a regular plankton flow from the south, shielded from sunlight by low-visibility inshore water, this deep coast community has a unique habitat, which is so far from the sun no plant life can compete. All these filter-feeding animal colonies have found a rare and perfect niche. Their diverse, quiet community is very stable and long lived. Massive sponges survive for around 100 years. Apart from the nibblings of triggerfish and bacterial attacks that may thwart their chemical defences, there is relatively little disturbance down here. Competition for standing room on the limited area of rock is the main pressure on life.

Then, at the end of my brief sojourn, I gasp in amazement: fringing a more elevated rock, a thicket of lolly-pink and ivory-white branching corals, some as wide as my forearm, deep-water relatives of the stony corals that form reefs in the tropics. All too soon my fins must drive me skywards, head buzzing with surprise and delight.

UP THE SLOPE

Now, towards the headland, I make a planned series of photo dives to see how seascapes change; how the interplay of light intensity, rock angle and water

Up the undersea slope: two red moki *Cheilodactylus spectabilis* in a cave.

motion determine the living conditions of each habitat as I ascend a giant staircase towards the tideline.

From the deep outcrops at 45 m I fin shoreward across a white sandy plain that rises gradually, rippled by the constant inshore current. At 30 m I encounter a region of low rock patches: a platform reef from the ancient coast now partly buried in sand. This is the beginning of the headland slope. It is an alien, faery world, a fantasy garden of erect sponges of every shape and hue but seldom symmetrical like those deeper down. Even at this depth storm swells have some influence.

Further up the slope at 24 m, I find tumbled boulders and entranceways to winding, gravel-floored canyons, deep trenches and narrow fissures – formerly wave-gushing surge channels of the ice age.

As I ascend a tilting rock corridor the twin walls become more and more profusely encrusted with filter-feeding animals: flask sponges, soft corals, pencil-shaped bryozoans, patches of waving orange and white anemones, seasquirts and weed-grazing tiger shells. Polka-dotted orange-and-white nudibranch slugs munch grey sponge tissue and bryozoans.

From narrow fissures forests of rock lobster feelers thrust. In the shadows beneath ledges night-feeding fishes dash to and fro in their day dormitories.

Outside the open reef slope is sprinkled with the first of the kelp plants, from small, stunted wisps to drooping tresses on long spindly stalks, all struggling for light. But from 18 m upwards, the kelp forest thickens until it forms a

continuous swaying canopy that shades the rock beneath, favouring certain delicate forms of encrusting creatures.

As the influence of wave surge increases, the canyon walls become even more richly colonised with plankton-feeding animals especially adapted to the hurly-burly of upper level life, each accurately oriented at right angles to the vigorous seesaw water motion up and down the hallway: *Solanderia* sea fans, with a delicate white tracery of tiny stinging polyps supported by a horny brown skeleton that recalls the clifftop pohutukawa trees. And huge mauve tube sponges swaying on a base corrugated with wrinkles for flexibility.

Now the reef fish community is close to its best. Beneath jumbled rocks tribes of blubbery-lipped red moki hover; wrasses scull about, triggerfish nibble sponges, goatfish probe all surfaces with their sensitive barbels; weed-grazing parore and silver drummer dart here and there like frightened sheep.

The upper rock slope has become a complex city luxuriantly carpeted with red, green and brown seaweeds, a downtown maze of tunnels and submerged skyscrapers, with avenues, malls and arcades. Wherever violent water movement is buffered, shielded or contained by large rock outcrops, trenches or overhangs, feeding conditions become optimal for encrusting communities and the concentrations of reef fish are greatest.

Finally, within 9 m of the tide edge I am jostled to and fro over areas of bare rock deeply dissected with guts and trenches. All the flat surfaces bristle with spiny armies of sea urchins, voracious five-toothed weed grazers. Up here the

In the shallow zone: male butterfish *Odax pullus* with long trailing fins like seaweed.

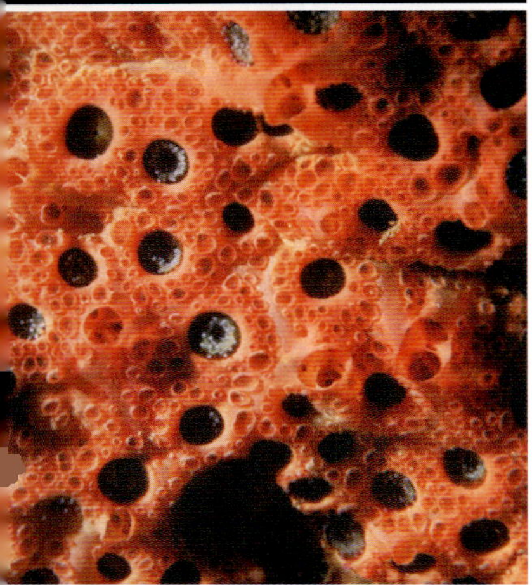

ocean's thresh and ply is seldom still and life is tougher. Diversity is reduced, but there is plenty of light for the seaweeds and with the high energy of water in constant motion, food is abundant for all the clinging herbivorous creatures that can withstand the washing-machine tumult: paua, chitons, topshells, limpets, sea snails along with predatory starfish and trumpet shells.

Whenever the rock angle is too steep for weed-grazing urchins to cling during storms, the seaweeds are especially luxuriant. Rock walls near the bottom of the tide become vertical jungles. The interface is fringed with clumps of wiry-leaved strap kelp, tangled thickets of bladder kelp, curtains of leathery herringbone weed, rococo-patterned weed and emerald ruffles of fast-growing sea lettuce – all surge-loving plants. Swirling and melting among the golden fronds are weed-eating fishes: butterfish with long trailing fins and frenetic marblefish, like panicky maiden aunts, disturbed by my bubbles as I break through to the glaring light of day.

WHERE WORLDS COLLIDE

On a day of powder-blue calm I track along the beach past forests of birdsong to a platform reef at the cliff base. It is almost low tide. Bracing whiffs of iodine. Warmth radiating from an expanse of jumbled grey rock. Apart from gurgling and trickling sounds all is quiet along the sea margin under the low noonday sun.

The entire region between the tides is really a savannah of rapid-growing microplants sprouting over the rocks, top-dressed twice daily with nutrients from the sea. At high water a few hours earlier, armoured herds of molluscs, limpets, chitons and snails were grazing a thin photosynthetic film like the fuzz on a peach. Passive plankton strainers, barnacles and tubeworms were netting their prey. Carnivores roamed among them, all with armour-piercing weaponry.

As I set up my camera gear I recall stormy days when I've watched from the clifftops as huge breakers smashed to smithereens over these same reef flats. And I think of that elegant sponge garden 10 storeys below the surface – a dim, tranquil, tideless world where temperatures change about 8°C seasonally and space is scarce. But from tideline to cliff face I am viewing *a battlefield*. Vacant lots abound, for which the asking price is exorbitant. Here temperatures soar and plummet some 50°C. Salinity can change abruptly with deluging rain or searing evaporation. Extremes of wave violence, desiccation and tidal change make this an unpredictable habitat. A zone of attrition where just a few highly adapted, risk-taking pioneers with armour, cement or powerful sucker pads can meet the demands. The winners have little competition and are in great abundance. Their populations are short-lived and can crash very suddenly.

I begin my camera journey from tide margin to cliff by exploring the tide pools. Some headlands rise sheer from the sea with no platform at all. On such promontories the varying degrees of exposure are compressed dramatically: the gradient of dynamic forces is so steep that life patterns appear as a sequence of distinct colour zones stepping up from the brown band of rubbery plants through mauve encrustation to barnacles and lichens of the splash zone. Tide pools are microcosms of the inshore marine world – an opportunity for the non-diver to enter the ocean. Pools at the lowest levels offer the greatest abundance and diversity.

From vertical, weed-fringed walls I peer down into a crystal chamber, a reflection of the food web in the sea. On the floor a large green anemone awaits unwary prey with swaying tentacles of doom. A well-camouflaged crab scuttles from beneath a boulder. An assortment of battered-looking shells totter and waddle about – the homes of scavenging hermit crabs. From a strange hairy shell a big leopard spotted snail extends its siphon, tasting the water, on the hunt with its rapier-like radula. Sinuous arms of brittlestars wave from a crevice. Horror story: an eleven-armed starfish has engulfed a small sea urchin. The spiny sphere is clutched to its mouth. The star's stomach is everted, digesting its living victim externally. Perhaps next on the menu, a cat's eye snail with pearly patches on its shell grazes a delicate fuzz of green weed. A sable rockfish spurts to a crevice. A clutch of eggs has to be protected: tiny sago-like translucent spheres within which eyes and tails squirm and pulse, soon to emerge as larval fishes at the mercy of a million mouths. Plant eaters, scavengers, carnivores – a cross section of the deeper world beyond.

Towards the cliffs the rock platform rises gradually and these upper pools become smaller, brackish and less diverse. Beyond the pools the low-tide battle for survival in air is in full swing. Clamped tightly to its home base the limpet shell no longer glides over the submerged rocks on its sucker-pad foot, head swinging in arcs as its radula shaves the microplants. Really it is a primitive snail, with a streamlined, flattened cone that copes with severe turbulence and hot sun. I find one under attack by a big white rock whelk.

In crannies and crevices, clustered into dense huddles to retain moisture, are the jet-black domes of herbivorous sea snails. Further up, the rocks are blanketed with the white pillbox houses of barnacles. It is hard to believe this limestone cup houses a minute shrimp-like creature, a crustacean that has spurned mobility for a passive, plankton-straining lifestyle. The barnacle awaits the few hours of immersion twice a day when the four panels on its roof will slide ajar, enabling six feathery legs like eyelashes to emerge, flicking and withdrawing rhythmically.

OPPOSITE, TOP TO BOTTOM: The shallows and intertidal zones, rock pools and tidal flats: New Zealand half-crab *Petrolisthes elongatus;* colonial coral *Culicia rubeola* on red sponge; hydroid fan *Halecium beanii;* waratah anemones *Actinia tenebrosa.*

LICHENS: AN ANCIENT PARTNERSHIP

Algal cells that live in symbiosis with a fungus, lichens are pioneers of the harshest places on earth: mountaintops and tundra, Antarctic rock, tree trunks and peat bogs. We may one day cultivate lichen on Mars! Although fungi live on dead plant matter they cannot use sunlight to make food, as do the algae. But fungi are able to glue themselves to bare rock, providing a good home for the algae colony living in their tissues. This coalition produces a lichen. In a parallel symbiosis the reef-building corals have been discovered to depend on algae in their tissues for nutrients – as do we humans on our gut flora.

A larger species, higher on the shore, just spreads its long curled, hairy legs across the prevailing current like a net. During spells of calm weather these yellow-and-grey acorn barnacles, packed tightly in a honeycomb pavement, may not be wetted for days at a time.

Patches of bivalve oysters share the active plankton-feeding lifestyle. With one valve cemented tightly to the rock, the other maintaining a tight moisture seal, the rock oysters await the rising tide, when their siphon tubes will emerge to suck in a nutrient stream of plankton and oxygen. Along with the oysters, black mussels and tubeworms cling to vertical surfaces, filter feeding in a similar way when the ocean returns.

Beyond the reach of the tide, clinging to the cliff face in seams and crevices, are the spiral-coiled, blue banded, sharply conical periwinkles like tiny seeds. Such fragile creatures avoid the wave violence by living so high up they are virtually land dwellers, grazing lichen and sparse blue-green algae in the splash zone.

In this short journey from ocean brink to splash zone I can explore a range of habitat changes equivalent to the ascent of a 3000-m mountain – or from tropical rainforest to Arctic tundra. The zoned seashore on a headland coast is one of the most concentrated, regularly patterned and ancient communities on the planet.

UP THE CLIFF AND INTO THE FOREST

What life is there up on the weathered headland ramparts? Camera dangling dangerously, I scramble for finger and foot holds. Above the last grey bands of jagged barnacles I find the spray-damp rock face becomes slippery and sooty black. A dark film of gelatinous microplants spreads across the fractured strata, grazed here and there by pockets of periwinkles. This maritime splash zone is one of the planet's least hospitable habitats and it extends near sea level all around the world. The tarry lichens that live here emerged from the sea aeons ago. They are much older than the rocks they encrust.

Further up I reach a zone of crusty lichens. They are deep orange or yellow, felty roughenings in patches, rosettes and wavy-edged discs. The bright pigment is a protective filter to reduce ultraviolet light under intense exposure. Lichens have the capacity to dry out to a crust and rejuvenate when moisture becomes available.

A third zone develops: sage-green or pale-grey lichens with brittle, twisted edges and abstract patternings. Down the rock face a red weep of water trickles in an elaborate series of flutings. Calcareous deposits, each a minute demi-basin, a hydroponic pool of rusty-brown algae. To my surprise I find black sea snails feeding on it. This algal soup must be nutritious for them to venture so far from the sea. Tunnelweb spiders lurk under ledges nearby.

Then the first flowering plant, a rock lily with sword-like leaves. In summer it produces long sprays of sweet creamy blossoms, followed by wine-red berries.

Wherever the cliff offers a small advantage, a fissure or cranny, thick-leaved plants sprout; plants with specialised strategies to conserve moisture. Some minimise water loss by only opening their pores to breathe at night.

I find pink-flowering ice plants, leathery glassworts, heart-leaved native spinach and finger ferns. Their leaves have thick, waxy cuticles or dense water-storage tissues designed to withstand extreme desiccation and blistering midsummer days.

Beyond my reach the windswept heights and eroding brink are sprigged with spindly, twiggy shrubs, prostrate ground huggers, such as the mirror leaf taupata, a coprosma related to coffee. I see springy cushions of sweet-scented mingimingi and thin stems of potato orchid.

Thrusting out impossibly from the precipice edge, supported by iron-hard, gnarled and writhing root systems, which secure and bond from erosion the last relicts of soil cover, are the kings of cliff dwellers, the gale-defying, enduring pohutukawa trees.

Beyond, a green diversity invites me to explore further. Glimpses of shiny flax leaves, crests of puriri, karaka and ngaio; nikau palms, tree ferns and golden tassels of kowhai – the hinterland beckons.

OPPOSITE: Up the headland cliff: golden shield lichen *Xanthoria parietina*.

TOP TO BOTTOM: Calcareous water weep; rock lily wharawhara *Astelia banksii*.

TOP TO BOTTOM: Whale Bay Beach; Cabbage Tree Bay.

CHAPTER 4:

A YEAR IN A HEADLAND FOREST

As much as I learnt from my day on the coast, it was nothing compared to the secrets that were revealed to me and to Jan by patient, methodical study through the changing seasons of the year. Month after month, tree by tree, we recorded the forest cycle, marvelling at its complexities – and at how little one can gain from a casual visit to the coast unless a special effort is made to include the time dimension.

SPRING

On a warm spring morning I wind my way down the cliff path to Whale Bay Beach. In this sweet-scented forest from late August a festive time begins. I pass the hollow puriri tree near the entrance stile. By the picnic table with its panoramic coastal view I hear a plosive cluck and a chortle of tui song. A drunk on the nectar trail. A thick whirring of wings and a black ellipse hurtles from the cliff-edge kowhai flowers and swoops down the path towards the next golden grove. I follow, eager to catch a tui binge on film.

My way leads through a glade of broadleaf trees with understorey shrubs, scramblers and vines. The canopy trees are glossy-leaved karaka, taraire and puriri with red matipo, rewarewa and needle-leaved totara; most of them are berry producers beloved by the headland birds because these trees spread their bounty through the year. Forest margin trees and shrubs add to the richness: fragrant mingimingi with delicate panicles of miniature white bells; manuka bushes white with snowflake blooms; tufty-headed cabbage trees like giant stalked lilies and broad-fronded nikau palms, relatives of the coconut.

Crowning the second knoll are three immense, sprawling pohutukawa, each supporting masses of epiphytes: fan-shaped perching lilies like big birds' nests and clinging rata vines that snake their way up the host trunk. I try to visualise a northerly storm when those great drooping boughs, almost scooping the ground, would flex and thrust, thrust into the teeth of the gale. Clifftop branches must bear enormous strain, far beyond the capacity of hammered steel.

The path dips to a narrow saddle flanked with glimpses of shimmering bays. Now the way is hedged with wind-compacted tree daisies, hebes, white rata

New Zealand pigeon *Hemiphaga novaeseelandiae* in kowhai.

bushes with their shiny masses of disc-shaped leaves flush with new growth, gathering strength for end-of-year flowering profusion.

The path skirts a steep, moist gully plunging to the sea. In this cool dell the rainforest is dense, diverse and subtropical. Giant hollow puriri lean like Pisa towers and the smooth green trunks of nikau palms glisten with beads of dew. One extends a large spray of fleshy pink flowers, recently burst from a papery sheath, and abuzz with black native bees. Every glance yields diverse textures and complex patterns of foliage, bark, lichens, mosses and entwining vines. I find every possible variety of leaf shape: narrow and serrated; broad and crinkly; heart shapes, ovals, five fingers, thick and leathery; thin and papery; broad ferny feathers; needles, spears and swords. Each is an answer to the plant's basic needs: to gather sunlight and exchange gases and water vapour.

Suddenly I am there. A divergent path leads me up to a grove of kowhai trees. A New Zealand pigeon or kereru perched on high flicks its head inquisitively. A red-ringed eye gazes about. Emerald and mauve neck feathers catch the light like shot silk. But the world's largest pigeon does not flee. It continues to pluck the

A tui *Prosthemadera novaeseelandiae* revelling in kowhai.

DIVERSIFY AND PROSPER

In the undersea world 10 storeys below this headland, there is a diversity of widely divergent life forms – sessile marine animal colonies variously structured to filter, snatch or sting plankton food, prepackaged bundles of energy, passing by on the ocean current.

Now, in an air world, the needs are different: this community of plants competes for nutrients, moisture and sunlight by spiralling its root systems deep into clay and rock. Its energy comes from the sun. It needs animals, sometimes the wind, to disperse its seeds and sex products. But why such a diversity of leaf forms to perform the same task? Certain designs would best conserve water, resist high winds and salt spray, or gather light more efficiently in the shady understorey. But that does not explain the rainforest's endless variety of leaf patterns. Scientists now suspect this may be a response to the aggressor. There may be a pressure on trees to become diverse in lifestyle, shape and chemistry so that no single disease organism can attack them all. Diversity keeps the enemy in low numbers and diseases attack the most common species. Ironically, the diversity of rainforest may depend on a healthy crop of diseases! No wonder I find such variance on this Tutukaka coast headland. The rainforest of New Zealand is an assembly of species related to trees in New Guinea, New Caledonia, Australia and South America, descendants of Gondwanan ancestors.

new-growth kowhai leaves.

Whirring wings and a courting tui alights. Nearby females or rivals, with gleaming blue-black feathers and white bibs, hop amidst the yellow blooms. A sleek head twists to thrust its curved beak up into a dangling trumpet, drawing out a tot of mildly alcoholic nectar. Long stamens inside the flower gild the bird's head with pollen grains.

Along the pathway a shrub takes my eye. Clusters of perfect bell-shaped pink flowers with lovely curving lips. A rare *Pittosporum umbellatum* tree I had often passed without noticing. Down on the sandy beach I find big elongated purple-red berries of tawapou, dropped by pigeons in their plunder. This is a rare coastal tree now endangered.

A month later when I revisit it seems the headland is celebrating a mass marriage. Garlanding the canopy there is now is a bridal finery of white, lemon-centred stars. Each male clematis bloom is an elegant diadem. In more advanced female flowers this has opened into a whorl of gossamer-winged seeds. I call them the *fuzzies*. Bees tumble and hover in an orgy of mutualism.

I reflect: rata vines and bush lawyer, clematis, pandanus (kiekie) and supplejack are all vines or lianas. In a rainforest with so many plants competing for space, sunlight is at a premium. Getting up into the canopy and gaining enough light is a real battle. The vines short-circuit the problem. They don't waste time and energy producing a big trunk. Their tendrils just run up others and spread. Some, like the rata, the puka and the strangler fig in the tropics, *may* completely take over the host tree and replace it with a stout trunk.

In shadier glades I find the heart-leaved kawakawa bushes have erected on their zigzag black branches further decor: spiky male catkins like green candelabra. Each column is elaborately patterned with the double spirals of the Fibonacci series, the helical growth pattern that so permeates nature. I see it under the sea on certain *Steganoporella* stick bryozoans.

A fantail twitters downscale. Launching from a twig it flutters and dives, beak clacking as it devours midges in mid-flight. I recall the undersea cliff where V-tailed demoiselles are darting and pivoting, snapping up plankton creatures on the current. Over the coming summer both will nest and hatch young.

A trill, then an ear-charming three and a half bar melody. A grey warbler's courtship song. Over every branch and twig in a manuka tree it flutters and scrambles, minute ruby eyes searching for grubs and bugs. An effective plant cleanser. Over the reef a wrasse sculls, fossicking with steady eyes for tiny edible creatures. If a reef fish has a parasite infection this wrasse will peek and pry, removing lice from gills and fin bases. A cleanerfish service. At dusk the morepork, ruru, flits through the trees like a whisper, hunting the first insects to stir abroad. Below the waves with the fading light a toadstool grouper darts over the bottom and pounces, snapping up the first crabs to venture out under the cover of darkness. A dagger-beaked kingfisher is motionless on a branch ready for any prey that moves. Mid-water a John dory hovers like a wraith, its vacuum-cleaner mouth almost invisible to demoiselles that stray past. Similar in form and function, gape-jawed lizards and lizardfish lurk on forest floor and reef base, ready to seize any small creature that blunders by. A diver sits amidst the trees and dreams of two worlds that are really one.

SUMMER

Over the months of October and November the coastal forests burst with flowers: among them karaka, tarata, karo, kohuhu, kanuka, miro and hangehange; kiekie, the pandanus climber; bush lawyer and the world's biggest fuchsia, the papery-barked tree fuchsia with its rare blue pollen, prelude to summer's sweet red konini berries.

During November a giant green moth with golden markings clings to tree trunks so resembling a leaf as to defy perception. This is the parent of a grub that devours the interior of puriri trees without destroying its host, like the sentinel tree by the gate that I love to crawl inside and from there scare my visitors. That

OPPOSITE: Tawapou berries *Pouteria costata*.

TOP TO BOTTOM: Fibonacci spiral growth patterns: kawakawa *Macropiper excelsum* flower column; stick bryozoan *Steganoporella neozelanica*.

vast hollow tree still flourishes on the windy ridge. The morepork feasts on these moths in flight. And uses the hollow trees for nesting.

December is climactic – all along the headland coast, as if to celebrate the solstice, crimson pohutukawa blossoms set the forest ablaze. Their stamens redden the beaches and pathways, and all the nectar feeders embark on an orgy.

Because the birds and insects of a rainforest have continuing needs and the trees seek to exploit their services to the utmost, flowering continues. During high summer towai, white rata, matipo and cabbage trees flower along with perching lilies and native blueberries.

THE COOLER SEASONS

But now the autumn forest begins to offer other food: as lacebark trees bloom the berries of karaka, karamu, nikau and cabbage tree begin to ripen. By May kohekohe and taraire fruit are delighting the pigeons. And even in the coldest months white sprays of hebe and pink puriri flowers continue to unfold.

In the very dead of winter certain processes begin beneath the forest floor. Complex networks of fungal threads start sending up their fruiting bodies. Food storage depots are activated and orchid bulbs begin to stir. Increasingly from July, surprises greet me. From rotting logs and damp places a weird array of toadstools and other fungi start to sprout: the fly-duping basket fungus; the stinkhorn that so resembles a sea anemone; the bird's-nest fungus with its tiny yellow cups containing egg-like spores.

POHUTUKAWA PORTRAIT

The same diversity I love so much beneath the sea was all around me in the coastal forest, but much more subtle. When they bloom, plants suddenly reveal their individuality. And in the rainforest this process is staggered throughout the year. As when underwater I soon found that regular visits to the headland rewarded me with the totally unexpected. Just when I thought a scene was familiar to me, I would get a jolt. Why couldn't I recognise a new tree simply from its foliage? Its flowers so enhance a plant's individuality its recognition can be rather like a human face. When I decided to follow one tree intensively for a year, the reward was immense. I came to see each plant in a new light. Considered as a cyclic process rather than a static entity, the *persona* of each tree emerges. Then, by a quantum leap, the diversity of forest lifestyles expands.

OPPOSITE, TOP TO BOTTOM: Tree fuchsia *Fuchsia excorticata*; White rata *Metrosideros perforata*; dripping cap toadstools *Mycena austrororida*.

ABOVE: Sandy Bay pohutukawa.

In springtime the pohutukawa starts unfolding new shoots from twigs where last year's flower heads still retain a few fibres in their upturned seed cups; still awaiting one more storm to buffet the branches and shake them loose into the wind and the wetness. Other branches develop special bud arrays that swell into silvery green capsules, each with a hint of red on the crest. My eye traces the symmetry underlying the bud cluster: an exquisite strategy for maximising attractiveness for the pollinators.

By early December the buds are swelling. Pale yellow sheaths slip away. Then a fountain of red stamens bursts forth around the margin of a glistening emerald lake. A quivering palisade of spears tipped with pollen-shedding anthers. Beneath this lake lies a gentle furry dome. From its centre a hollow column rises: the female, pollen-seeking stigma. Over the days I notice that the flowers on a cluster do not all open at once. And each green lake only brims with nectar for one day. This is a clever strategy to extend the flowering period so that congregating birds and insects can cope with the nectar flow and best share their pollinating services among the community. Looping up the silvery-fuzzed stem, a red grub no longer matches its food source as it seeks another bloom.

During ensuing weeks the red splendour of coastal pohutukawa trees diminishes until by February the flowers have all been replaced by seed clusters. From each green lake a silvery dome expands, rising as a hemisphere from the ovary. Three fissures split the cupola. Green turns to brown. The dome roof ruptures and three curving pouches packed with honey-brown fibres face the sky. Why, I wondered, are all the seed cups directed upwards, instead of scattering their contents from an inverted cup?

Over the winter I learnt the answer – yet another clever plant strategy. Only when conditions for seed dispersal are ideal in this harsh cliff-edge environment, when rain and high winds are lashing the forest, would the seed cups toss and plunge violently enough to spill their contents; all their tiny bits of fibre rather like babies' eyelashes. From crevices in rocks or clay, in full sunshine a tiny root unfolds and a simple pair of green leaves seeks the lengthening days.

POHUTUKAWA UP CLOSE

I thought I had some understanding of the pohutukawa. Then the earth shifted under me: I saw a sequence of pohutukawa photographs made with an electron microscope. I went on a journey into the leaf of a plant.

On the fawny underside of a leathery ellipse, veined like a feather, I see a tangle of densely matted, pale-brown hairs. Magnified 700 times they resemble a mass of coiled, squirmy worms, several layers thick. Each one twists like a fibrous rope, tapered at the end. At the base are rows of domes like gun turrets on a bomber. Each has an oval gap on top. The dense felt of hairs restricts water loss from pores in the leaves.

Magnified one thousand times, the dome or stoma resembles an inflatable igloo. I imagine I can enter the cuticle of the leaf through the portal of the stoma, passing through an antechamber and a pair of guard cells, water-inflated collars that can expand to close the stoma – like the airlock of a space station. This hydraulic valve controls the entry of gas and the exit of water into the leaf when the plant needs protection. Now I can wander freely through a system of passages like the interior of a castle built with hexagonal bricks. Each moist green tile is the wall of a cell within which the leaf factory is at work. Water and gas permeates these walls. Sun-powered chloroplasts are transforming chemicals into plant food – the basis of life.

With a further effort of imagination I journey from this leaf factory to the soil, following pathways of sap running beneath the tree bark, all the way down to the root hairs, spiralling deep inside the headland, where nutrients and water are entering the tree, rising again to the growing leaves and blowing away as vapour on the sea breeze.

OPPOSITE, TOP TO BOTTOM: Pohutukawa year cycle portrait: new leaves; buds; blooms; seed capsules.

ABOVE: Pohutukawa nectar in hexagonal emerald lake.

THE PATTERN THAT CONNECTS

When I consider the overall picture: the whole coastal headland 'slice' in profile, as in a side-on diagram, from the undersea world up through the interface to the rainforest promontory, major questions arise. Why such diversity above and below? Why doesn't one organism, more successful than all the others, take over and dominate? How is this diversity maintained through time?

The answers prove to be as complex as the worlds I am exploring, but certain principles stand out, forming a pattern that links both worlds – an underlying blueprint. On the undersea cliff and in the headland forest there are key animals that act as the agents of diversity. Disturbance is the random event that leads to such an exciting mosaic of interrelated life forms.

Head down, bulging eyes swivelled forward, diamond body vertical and tail waving to give purchase to its powerful jaws, the triggerfish nips at a patch of yellow sponge. This fish can feed on almost any living thing. In one area every square metre of reef surface received 150 bites each day. The toughest foods, neglected by most reef fishes, are delicatessen for the triggerfishes: sponges, bryozoans, polyps, tubeworms, barnacles, urchins. The local triggerfish tribe recycles a fifth of the entire rock surface each year.

Removal of such an astonishing number of food pieces annually has a profound influence on the diversity of animals in an encrusting community and helps maintain its ecological balance. Like the wrecker's ball on city landscapes, any source of disturbance, whether from storms, bacterial attack or reef grazers, is essential for the maintenance of diversity.

Sea creatures have staggered breeding times. Any month can be spawning time for one organism or another. Whenever a new surface is laid bare, be it January or June, a different coloniser, voyaging in larval form, will settle out of the plankton on to the new piece of real estate. And so, the patchwork quilt diversity of the undersea headland is maintained.

Above in the headland rainforest there are no ocean currents to carry away seed and wind is not always effective. Tiny seeds have limited food stores. Many of the headland trees rely on animals to distribute their pollen and offer them seed large enough to carry a good energy intake.

Autumn in the headland rainforest that I haunt: a heavy seed bounces off my scalp. I glance up. Perched on a branch overhead is one of the largest and most beautifully plumaged pigeons in the world. And it has just relieved its digestive system, performing its role as keystone animal. It is propagating the tataire tree in a random pattern through the forest. Soon a seedling will sprout and if a gap develops in the forest canopy from fire, wood rot disease, heavy wind or rain, the pigeon's work will result in a replacement tree.

Animals replace ocean currents as distributors of pollen and seed. So in this forest it is storms, lightning and diseases that are the major disturbers. The outcome, in both worlds, is diversity. Without disturbances and key animals a few species could dominate, overwhelming the others, and then succumb to disease or insect attack. The ecological security of diversity would diminish. Viva the triggerfish and the native pigeon!

But I am sobered when I consider that both animals had no major predator until humans arrived on the scene (and they became endangered). Accordingly the pigeon produces just one egg every year or so. The well-armed triggerfish is a nest layer with relatively few young. Their low fecundity is adapted to their replacement needs: survival chances were once high as so few predators could attack them. Then they became *creamfish* on our dinner tables!

What is the future for keystone animals and the worlds they influence? As catalysts for change so much depends on them. They are so fragile in our hands. Humans once shared this role: pollinators and seed spreaders of Eden. Do modern humans have an ecological niche, I wonder?

TOP TO BOTTOM: Keystone animals: A pair of triggerfish or 'leatherjackets' *Meuschenia scaber* grazing a rock surface; New Zealand pigeon feeding on matipo berries.

Our tiny cottage on the mangrove bend we studied by day and night.
COLIN GANS

CHAPTER 5:

A MANGROVE WORLD

After several years exploring coral reefs aboard Dr Walter Starck's undersea research vessel *El Torito*, an intensive period of ecological study had been completed. In 1974 we returned to live in a small cottage beside the Matapouri mangrove river. I recalled Walt's words about mangroves during cabin talk: 'I've always had a fondness for mangroves,' said the scientist-philosopher, who grew up around them in the Florida Keys. 'They're one of the places where life has reached its fullest expression; among the most productive communities on the planet. The ultimate evergreens.'

Aroused by coral reef communities and tropical rainforest, my passion for biodiversity spilled over into the mangroves at my door. Walt's words kept coming to mind. Why did he value them so highly? Why were they so incredibly productive? To my delight I began to find a complexity of life in a world of mud and flooded forest that matched the tropical reefs I still yearned for.

ABOVE: First bend at low tide.

OPPOSITE, TOP TO BOTTOM: Like a miniature glass helicopter a translucent prawn *Palaemon affinis* hovers; a cat's eye sea snail *Turbo smaragdus*, a herbivore, grazes algal fuzz on a mangrove breathing root; dorsal fin signal flag erect, a male triplefin *Forsterygion varium*.

I built an observation platform above a winding river glade; I suspended myself face-down above crab city in a rectangular net; I crawled beneath overhanging mud banks with a torch and high-power lens; by day and night I dived with scuba; on a special surfboard with a window-panel I drifted upstream silently with the tide into this hidden world.

At a time of exceptionally heavy rainfall in 1975 I saw all the sand swept away from the roots of two lone mangrove trees near the mouth of the Matapouri River. The entire root system was exposed – an elaborate radial web of cable roots and risers that extended beyond the height of each tree. In more typical situations the interlaced root mass may be up to five times the diameter of the tree's canopy. There is no taproot. This massive raft of roots anchors the tree against the most extreme erosive forces of storm energy, wind and water pressure, effectively consolidating the surrounding area.

DIVER'S BEND

On the first bend upstream, I discovered my favourite diving area. At this point the river is still a broad estuary, a sandy plain crowded with large, spreading mangroves. The streambed is a firm expanse of cockles and empty shells. At high tide the water here can be very clear. During springtime coastal waters often become green with plant plankton. So effective is the filtration effect of all the plankton-sieving life forms in the mangroves that the water is often much clearer when the tide recedes than when it came in.

On this bend of the river I can find most of the life forms in the mangrove community and in ideal conditions for photography. But to get close to some of the shyest I had to restore a rare piece of diving apparatus: the bubble-free, silent oxygen rebreather. My lungs expand and contract in rhythm with a rubber bladder strapped to my chest. Replenished from a small steel cylinder, warm, friendly oxygen is quietly cycled to and fro and not a single bubble escapes to alarm the tiny prawns and triplefin blennies close to my lens. I am weightless and feel very like a fish myself.

Like a miniature glass helicopter a translucent prawn hovers, backing, advancing, rising and falling around the pencil column of a densely encrusted breathing root that quivers in the current. The prawn tries to avoid me. Rows of swimmerets under its tail pulse in waves that reverse or arrest as it manoeuvres deftly. Barnacles, oysters and dwarf black mussels cling to this breathing root like occupants in a high-rise apartment building, feeding on the rich flow of plant plankton borne in by the tide. A fat cat's eye sea snail glides over them, its radula, like a child licking at an ice cream, rasping at the thin fuzz of algal film that coats every surface in the mangroves.

A triplefin blenny hops across the streambed

towards me. Its foremost dorsal fin is erect, a red splotched signalling flag. This is a male making a bold territorial claim to this huge monocular rubbery monster. Nearby I witness a tragic battle: a blenny has seized the leg of a crab in its jaws. Just like a shark it rotates to tear the limb loose. Other blennies seize the crab's flailing appendages and provide leverage for each other until their twisting action snaps the legs off, leaving the crab helpless to defend their attacks.

Beyond this first bend the river enters a more typical mangrove world, flowing between steep, deeply undercut banks of the thickest mud. Along this stretch I set my observation platform among overhanging branches and, nearby, my crab hide. I found that some fish from the sea rest up here beneath the banks overnight. Eels by the dozen prowl for crabs, and a snake eel as thick as my thumb but 2 m in length lurks in the mud, just its glossy teeth protruding to trap prey.

One day it occurred to me – the riverbank mud does not behave as mud should. I had watched strong currents swirl up and down the river at times of spring tide and storm flood. Once I even feared that our little cottage would be swept out to sea. But the banks remained unchanged. No tree was even displaced. How could normally colloidal mud resist such immense hydraulic pressure? To satisfy my curiosity I had to steel myself for a mucky ordeal.

At low tide with a powerful torch and a hand lens I crawled beneath an overhanging mud bank until I could scrutinise the walls of the deepest recesses. To my amazement I found that what I had considered mud was really an array of tiny stones and boulders of varying colours and composition, held in place by a filigree network of fine white threads like cotton. It was just like the system engineers develop: wire netting groynes packed with boulders, to protect bridge abutments from erosion. The delicate hair roots that nourish the mangroves entrap and retain sediment so firmly that riverbanks of soft mud maintain themselves for years with apparent permanence. For this reason mangrove-lined creeks remain deep and navigable far up into the forest. Because of their elaborate root systems, mud no longer oozes as mud should.

Snake eel *Ophisurus serpens*: another lurking predator.

VERBENAS OF THE SEA

I enjoy a stimulating cup of lemon verbena tea brewed from leaves of a small green shrub in our garden. And the pink flower clusters of the Mexican verbena bush have a sweet perfume. But I was surprised to find they are both related to mangroves! The mangrove living in New Zealand belongs to the verbena family. A cosmopolitan species, it also lives in Australia, Asia and South Africa. A verbena that lives between the tides.

The mangroves belong to the highest group of plants, the seed bearers, whose earliest forms probably developed on land but adapted themselves to a marine environment to escape competition with other plants. Certainly no plant threatens their dominance of the difficult world where shore and sea meet. To survive on anaerobic (oxygenless) marine sediment the mangrove puts down its grid of dense, fine roots beneath the mud.

From these roots arise breathing roots or pneumatophores. Anyone who has been floundering in a mangrove forest will remember the thickets of pencil-like roots sticking about 5 cm out of the mud. Nets get caught in them, and they are almost as uncomfortable to walk on as a bed of nails. But these breathing roots are the secret of the tree's survival in salt water and mud. If they are destroyed or smothered the tree will die.

A SEED SETS SAIL

At the height of the northern summer, the story of an individual mangrove tree begins. If you row quietly through the glades of grey-green and bronze colour leaves, soft plopping sounds come to the ear. These are the mangrove seedlings. When the tide ebbs, thousands of these buoyant seedlings drift out to sea, eventually to become stranded on another shore. If conditions are right the tiny root is quickly established in the mud and a leafy stalk is sent up. As it grows in a tangle of roots, all sorts of debris collect around it: driftwood, leaves, shells and sticks. Within this shelter crabs and worms take refuge. Molluscs and barnacles settle on the roots and branches. In a short time a complete community of plants and animals is established, all tied biologically to the mangrove ecosystem: truly a cradle of the sea.

On a low-tide gumboot journey I find a newly sprouted mangrove on its side rocking gently at the water's edge. Its four hook-like roots are extended. Two of

them have grasped the mud. Soon it will hoist itself upright. Sometimes a crab assists a seedling to get established by dragging it into a burrow to dine on the fuzz of filamentous algae colonising its surface.

The mangrove embryo actually germinates while still hanging on the tree. This characteristic of mangroves gives them a head start in a difficult environment. Technically it is not a seed, but a propagule: its yellowy-brown, velvety heart-shaped casing, which conceals a pair of fleshy-green, energy-storing cotyledons, contains just enough air to give buoyancy for a long journey. The seed has developed from a tiny four-petal flower, golden yellow and deliciously fragrant – like fruit salad, some say. Clusters of mangrove flowers attract bees in winter when other nectar supplies are scarce. The silvereye is also known to drink its fruity nectar. Mangrove honey has a special flavour.

In spring the seeds begin to swell, storing energy so the young plant can root and develop rapidly once it finds a likely habitat, sufficiently sheltered sand or mud. Around Christmas the drift-lines along Northland beaches are littered with the green jetsam of mangroves. Waterborne dispersal is so efficient that these seagoing offspring colonise every tidal inlet along New Zealand's northern coasts from Raglan Harbour in the west to Opotiki in the east. I have even found an ambitious young plant clinging to stones in a rocky nook at the Poor Knights Islands – 22 km offshore. Frost seems to be the limiting factor for mangroves and trees become progressively stunted towards the south until they reach less than a metre in height.

Only those seeds that are marooned in ideal situations survive the floods and storms of their first winter. Initially the young plant feeds from its own food store, but its tiny pair of leaves like green wings soon begin to do their work.

LEAF SCIENCE AND ROOT DYNAMICS

Up in the cool leafy canopy above Matapouri River, examining a leaf closely, I suddenly see a diamond glinting. My macro lens zeroes in on a perfect salt crystal, faceted geometrically like an expertly cut gemstone. A tiny pore or stoma in the leaf has secreted this unusually large piece of salt, assisted by certain atmospheric conditions on this calm, humid day.

I see the mangrove's leathery leaf in a new light: a water-cooled solar panel, absorbing carbon dioxide through valves in its surface and using the sun's energy to combine this gas, and nutrients supplied by its root system, into plant sugars that feed the tree. To avoid heat damage to its leaf factories the mangrove draws up salt water and the leaf transpires it as vapour through pores in sunken pits scattered among a tangled jungle of white matted hairs on the underside. This dense insulating felt minimises the desiccating effect of wind flow and intense solar radiation bouncing off the water surface. The leaf's upperside is waterproofed by a thick, waxy cuticle.

Twisted and gnarled, branching widely from a short stout trunk with grey furrowed bark, the mangrove tree defies the worst of storms. The wood, brown, durable and heavy, is composed of fibres arranged in layers that criss-cross diagonally to give enormous strength and a flexibility beyond that of hammered steel. In the far north mangrove trees reach eight metres in height – truly an estuarine forest. Some mariners have learnt that the upper reaches of mangrove forests are the very best place for small craft to shelter unscathed from the fury of cyclones.

As for the mangrove root systems, their special survival secrets are revealed at low tide. Most tree roots grow downwards. But living in deep, anaerobic mud and salt water demands a radical strategy. Firm anchorage and a good supply of oxygen are the mangrove's primary concerns. This tree requires no taproot to seek water. Like a huge saucer, a shallow system of interlaced lateral roots and strong bracing cables radiates from the trunk. From these laterals every few centimetres grey-green peg roots ascend up to 50 cm like a dense array of asparagus spears. Flexible and corky, these aerial roots or pneumatophores are studded with pores called lenticels, through which air enters the spongy interior at low tide and is diffused throughout the tree. All these vertical snorkels and the fibrous mats of fine white feeding roots collect silt, sediment and detritus, seaweed, sticks and leaves, all building up around the tree, a couple of centimetres each year, buffering it from erosion, breaking up any wave action, and weighting it down against the forces of the highest tide and the worst flood. No engineer could develop a more effective system for erosion control. In the Thames area civil engineers have transplanted mangroves for this very reason. Near the Matapouri river mouth I experimented with young trees in a sandy area where erosion was a problem. To assist them initially I planted them through heavy plastic netting and weighed it down. Before long the trees had developed their own anchorages.

Close to my tree platform I selected a stretch of river and studied it with the tide in and out, filming the process of tidal flux, and I kept up my observation for more than a year. I realised that if mangroves were permanently inundated, they would drown. Tidal ebb and flow is essential for oxygenation and nutrient supply.

OPPOSITE, TOP TO BOTTOM: A newly sprouted mangrove seed *Avicennia resinifera;* mangrove seeds developing on the tree; a fragrant, honey-producing mangrove flower; life begins for a ubiquitous mangrove seedling: first leaves unfurl.

ABOVE: Mangrove breathing roots, pneumatophores, entrap its drifting seeds.

WHEN A MANGROVE LEAF FALLS

When old leaves become excessively laden with stockpiled salt the tree just sheds them, with enormous advantages for all the low sly lives in the mud below. Studies show that up to nine tonnes of dry leaf litter per hectare fall annually.

Even before a leaf falls from its twig the failing plant factory is prepared for recycling: it has already been colonised by bacteria and fungi. Within 24 hours of hitting the mud a host of marine microbes invades through tiny cracks in its waxy cuticle. Fungal spores like tiny sperms with wriggling tails swim right into the leaf through its open stomata or pores. Then they whip out filaments that drill through cell walls to feed on the carbohydrates within. Now other single-cell animals and plants, called protozoa and bacteria, attack. The micro-algae and fungi form a slimy brown film over the leaf surface. This is food for minute marine worms such as nematodes and flatworms and crustaceans like copepods. Almost invisible, a flatworm or turbellarian, just a millimetre long and transparent, glides over the slime, changing shape as it moves, guided by two light-sensitive eye spots. It sucks in prey half its size: nematode worms and copepods. Such a creature is just a mobile digestive sac.

Small crabs, shrimps and amphipods begin gnawing at the leaf and it starts to break up into a slurry of tiny particles, a nutritious soup infested with microbial life and their valuable by-products. The leaf components are eaten and excreted by worms, tiny molluscs, shrimps and crabs; and because bacteria in the wastes of these animals can fix atmospheric nitrogen, the food value of this organic matter is raised as it passes up the food chain. Grey mullet feed directly on the leaf detritus, scooping up mouths of mud and filtering out the enriched brew of mangrove leaf particles and micro animals. A 20 cm mullet filters around 1500 g of sediment soup per day; 450 kg per year!

With such lush, well-nourished, regularly watered forest shedding vast amounts of organic matter, and with the nutrient content of all this detritus being enhanced by all the decomposing organisms, an immense and constant food supply is being transmitted through the food web. This is nature's breeder reactor with no radioactive downside.

Salt crystals secreted from a mangrove leaf pore.

MANGROVE MUD

An essential component in all animal and plant tissues is nitrogen. Yet only a few simple organisms can convert atmospheric nitrogen, which makes up 79 per cent of the air we breathe, into organic compounds. Certain blue-green algae and bacteria are the nitrogen fixers at the base of all life. Indeed, it is suspected that life actually evolved from these single-cell organisms.

Mangrove muds offer three important sites for nitrogen fixation. The first zone is the epiphytic blue-green algae attached to the breathing roots and submerged trunks and branches. Then there is the nitrogen fixation that takes place on the actual surface of the mud where leaves are decomposing. Thirdly, nitrogen is being fixed by very special bacteria that live 10–15 cm below the mud surface: anaerobic bacteria that can metabolise sulphur instead of oxygen. The amount of nitrogen fixed in the leaf litter alone is equivalent to that in a pine forest on land. But more important than the rate of nitrogen fixation is its export through tidal flushing to surrounding ecosystems and food chains where rates of organic matter production are much lower.

On a hot and windless afternoon I roam up the streambed. I find a sill on the streambed where water is flowing over a worm-riddled log. Plant plankton is accumulating in a yeasty foam. This, under magnification, resolves into galaxies of diatoms, a rich broth of food for all the filter-feeding invertebrates, oysters, mussels and barnacles that encrust the breathing roots and trunks: the first step up the marine food pyramid. Then I look closely at the log. A network of golden-brown tubes forms a dense interlaced carpet over its surface; a fast-growing algal mat that is sprouting rapidly on every firm, sunlit surface in the mangroves. This is the favourite food for the parore that come up here with the tide. Like tiny rakes its special three-pronged teeth scrape off this rich plant material, which is renewed after a few hours of sunshine.

At low tide festoons of Neptune's necklace seaweed (*Hormosira*) are draped among the palisades of breathing roots, trunks and branches. They resemble the interlocking poppet beads children play with: plump chocolate-brown spheres, water-filled to resist desiccation, linked by tapering stems and branching frequently, each branch terminating with a tiny delicate bead in the process of new growth. I find these *Hormosira* morsels are tender and taste rather like an olive. In ideal places this seaweed carpets every centimetre of the mud surface in areas where the breathing roots provide secure anchorage. Curiously, *Hormosira* weed living upstream in the mangroves is completely free to rise and fall with the tides. There are no roots at all and it can photosynthesise in air or water. In mangroves it can rely on plenty of hooks and snags and is seldom washed out to

sea. Beneath its sheltering spheres I find a microworld of crabs, snails and worms that enjoy all the abundant feeding surfaces and protection from predators.

Near the edge of the streambed and over raised sills of muddy sand I find luxuriant beds of eel grass – rich green blades like a sward of unmown lawn in spring. When the tide is flowing it ripples in the current and traps sediment. Now the tide has receded it lies in a dense protective thatch. Under the mud its complex fibrous root systems anchor and stabilise the streambed. The eel grass is another miniature jungle where many different sea snails, spire shaped or spiral orbs, are rasping off the minute fuzz of epiphytic algal growths that sprout all along the blades.

A study has shown that more than 30 species of molluscs, worms, crustaceans and other sea creatures shelter and feed in one eel-grass bed near Auckland. This makes eel-grass flats a popular dining area for estuary birds and for coastal fishes when the tide is high.

Eel grass, like mangroves, is highly productive: figures of 10 tonnes per acre have been calculated. In the United States a square mile of eel grass yielded 235 million prawns and 95 billion molluscs. Eventually eel-grass blades break off and decay into tiny particles fed on by bacteria and fungi. Teeming hordes of single-celled animals, protozoa, propelled by lashing whips and beating oar-like hairs, jerk along amidst the green slurry. These millimetre-long white specks and motes of life become further food for filter feeders, barnacles, bivalves, crustaceans, snails and worms and, then, the birds and fishes.

Sand flounder *Rhombosolea plebeia* shelters in eel grass.

ARMOURED HERBIVORES: THE SECOND STEP UP

If you remain stock still in the mangroves at low tide, not even blinking an eye, the mud surface will come alive. At every slanting crab hole an occupant emerges sideways: a small crab with a rectangular carapace, olive green or reddish brown, and prominent stalked eyes. So visually acute are these crabs to the slightest movement that even a blowfly buzzing errantly over the pockmarked terrain leaves a metre-wide wake of abandoned entrances. Obviously survival depends on vigilance and unquestioning retreat from the least movement, rather than a risky pause for danger assessment.

Up in the canopy I rigged a rectangle of netting between the trees like a hammock so that I could lie prone and observe a broad expanse of mudflat undetected. From my godlike vantage point I overlooked a busy city. Each burrow is the centre of a crab existence. I counted more than 30 holes within a square metre. Individuals seldom ventured more than a few centimetres from their refuges. As soon as one approached another's abode there would be a ritualised threat display of wicked-looking, jagged-edge pincers. The invader would usually retreat. At peak breeding times (autumn and spring) aggression is

Tunnelling mud crab *Helice crassa*.

TOP TO BOTTOM: Mud crab *Helice crassa;* underside of a female with eggs; the formidable purple shore crab *Leptograpsus variegatus.*

more frequent and actual physical conflict may develop. But, for the most part, the crabs, like many other territorial animals, prefer to use threat gestures rather than risk a physical injury that could hinder survival.

With such a density of animals feeding on the mud every minute the tide is out, night and day, it must be very fertile. But what were they eating? I built a crab hide that enabled me to lie prone at mud level, viewing them through a hole in a screen that concealed my movements. I could thrust a tele lens through the opening and zero in upon a single crab burrow. Within seconds I was watching an armour-plated monster at work. It was amazing!

The reference books told me they just spoon up the sediments with their chelae and ingest them. What I actually saw was much more complex. After each tidal change the crab has to clean out its burrow anew. A small mound of tailings is deposited on the doorstep. Now, in the mangroves any and every new surface exposed to the warm sun rapidly sprouts a 'lettuce patch' of fine algal growth. The crab fastidiously plucks delicate green filaments from its garden and munches them like a monkey eating peanuts.

For me this was quite an insight. All over the mangrove flats the crabs were miniature eco-farmers cultivating greenery within their tiny territories, sustaining themselves entirely from the solar energy fixed by the plants that grew there each sunny hour.

On one occasion I saw a crab set out on an epic journey from its hole. What ensued seems like an adventurous saga about armour-clad knights in a mediaeval story book. Everywhere the valiant warrior crab went it encountered fierce opposition and bold threat displays. An opponent would approach, nippers aloft and open, body raised on tiptoe. The two rivals would draw closer until they touched. Sometimes provokingly they might even blow bubbles in each other's faces. Then the invader would swiftly scuttle away. I never saw an actual fight, but other observers describe combat between rival males where one seized the other by a pincer, lifted him up bodily and thrust him down on his back. Sometimes limbs are torn off, but more usually the interloper is simply chased off. No doubt with such conflict the distances between burrows accurately indicated the area of photosynthetic activity necessary to support each crab with an adequate 'lettuce patch'.

When I found a female carrying caviar-like eggs beneath her tail I reflected that virtually every inch of mangrove mud houses a crab. I thought of the immense quantity of crab larvae this burrow-riddled mud city must release to the hungry sea. And how important the crabs must be to the diet of all the wading birds, kingfishers, flounders and eels. Besides the mud crabs most estuaries have several other crab species: one that feeds underwater at high tide; the tiny spider-like pillbox crab that hides under weed; and the formidable big purple *Leptograpsus variegatus* crab.

SEDIMENT SWALLOWERS

Gliding over soft mud in the upper reaches of the estuary is a model for future Mars travellers who will have to recycle all their wastes many times during their two-year voyage. *Amphibola,* the brown mud snail, is another eco-farmer. It swallows twice its own weight of mud every hour, along with myriad single-cell organisms, bacteria and diatoms that thrive on the warm, highly enriched surface. Behind the snail is a threadlike trail of wastes. Its coarse digestive process actually enriches the output with nitrogen and so many organisms survive passage through the mollusc to multiply rapidly that a mud snail population can consume its own wastes indefinitely: nature's miniature 'breeder reactors', densities of which can reach 500 per square metre. *Amphibola* is an air breather with a primitive lung and buries for protection at high tide. Experiment shows that it would actually drown if submerged too long. This sediment-swallowing lifestyle is shared by two smaller snails, the spire shell and a tiny conical snail.

Further down the estuary are the 'bird's foot' shells: those three-prong patterns in muddy sand so resembling the imprints of wading birds belie the feeding efforts of *Macomona,* the wedge shell. From its deep refuge this slender bivalve raises a 5-cm-long feeding siphon that furrows the low-tide sand with a vacuum-cleaner action to glean its riches.

The brown mud snail *Amphibola crenata* is an air breather with a primitive lung.

PREDATORS OF THE TIDE POOL

One morning I found a tide pool in the mangroves: an outcrop of rock amidst all the mud and branches. Here, with a firm base, I could observe and photograph many of the more widespread, hard to find components of the mangroves. Especially the next step up the food pyramid: the predators.

But first, like a green, multifaceted brooch surrounded in a decorative oval of snakeskin: a chiton clinging to the rock, moving out only at night to graze on algal film. In a quiet corner a tiny pale-green anemone: a maw surrounded by stinging tentacles. Then I discover the whelks – hyenas of the mangroves.

If I crush a tiny piece of flesh into the pool, suddenly periscopes erupt from the sediment. Like a submarine rising from the deeps a curving hull surges up from mud, flakes of ooze spilling down its bulbous bilges. With its sensitive proboscis extended from a groove in front of the shell each whelk tastes the scent of meat on faint currents circulating around the pool. From every direction they home in on their prey. Soon a horde of hungry snails covers the target like wolves on a deer carcase.

Nearby an explosive click of its outsize pistol-packing claw leads me to the snapping shrimp, a predator that lives in a burrow or under a chunk of wood. One of its nippers is almost as long as its carapace. On the movable part of the nippers is a peg-like projection, which fits into the fixed part. Closing suddenly, like the release of a gun hammer, the peg slams into the socket. High-speed photography shows this produces a water jet generating a gas bubble that bursts with a loud bang to stun prey.

Next, the fearsome mantis shrimp. This is an active night hunter with rapacious attack apparatus: twin claws with the hinging action of a penknife blade which the shrimp can open and shut with lightning speed. The powerful segmented tail and broad tail fan enable it to scull backwards with extraordinary speed in a peculiar looping motion. A rapid digger, the mantis shrimp lives in a deep hole up to a metre beneath the estuary sand, from which it makes its regular hunting forays. It can inflict a nasty wound on a human hand and can easily cut a snapping shrimp in half. It has acute vision and surprising manipulative abilities. Aquarium studies have shown it is much smarter than most crustaceans and can build quite complex galleries of stones.

OPPOSITE, TOP TO BOTTOM: A tiny pale-green anemone *Anthopleura inconspicua* lit up by flash. Symbiotic algae live in its tissues and provide nutrients — as with reef corals; acorn whelks *Cominella glandiformis* predating on cockle; the snapping shrimp *Alpheus novaezelandiae* lives in a burrow.

DAY DIVES

Before making a series of night-time explorations of the mangroves I find it best to do some daytime reconnoitres. I enter the river well upstream when the tide has less than an hour to peak. As I cruise over the riverbed into the current

BELOW: Wade takes photos: It is shallow but I need scuba to remain still and operate my cameras.

any disturbance my fins create will not affect visibility, especially important whenever I pause to take a photo.

I fin beside steep banks looking up to where slender tapering peg roots are silhouetted like stagshorn coral. Among the roots dense drifts of leaves lie entrapped with an occasional bright red one fallen from a pohutukawa. Going past on the current, tumbling over and over, a pair of green mangrove seeds. A fat cat's eye snail grazes along the skeleton branches of a forest tree washed in from beyond, its twigs draped in strands of seaweed from coastal reefs. Amidst a tangle of roots a tiny cloud of silvery sprats huddles for protection. Like eyelashes blinking the feeding cirri of barnacles flick in and out of their housings encrusting the column of a breathing root. Below them the valves of dwarf mussels and rock oysters are agape, filtering the water. Near an overhanging bank juvenile fishes of many species from the open coast dart and hover. For them life is much safer in this nursery than in the open sea where they would dodge quick jaws all day long. Scientific studies show that at least 30 species of fish use mangrove estuaries as a habitat for feeding, as a nursery, or as a migratory path between fresh water and ocean.

Triplefin blennies are everywhere leaping in spurts over the bottom and along the banks. Their population density rivals the crabs, but they can only live where there is a permanent flow. I watch them darting up to snap at tiny invertebrates borne by the current. Males have a signal flag emblazoned on their foremost dorsal fin. During territorial disputes they erect it to warn off encroaching neighbours. Propped on their pelvic fins like tiny fighter planes these plucky little fish even try to frighten me off with their aggressive displays. The same signal flag is used to court females. In winter the males will set up a nest site and guard a clutch of eggs until they hatch.

Where the current creates a surface vortex on a bend, a pool of brown scum collects: a rich soup of diatoms. In such places I see the blennies swimming upside down just beneath the surface, pectoral fins spread like fans skimming the scummy ceiling, mouths slurping at food items. Very adaptable feeders!

Another tiny predator is the rockfish. With its long, jagged dorsal fin it reminds me of a big grouper in miniature. I can never find it while diving, but if I turn over any log at low tide I usually hear a wriggling and a splashing. A rockfish for certain. I restore the log promptly. In winter the males guard eggs in the chambers of worm-riddled logs.

During my downstream journey I encounter timid schools of silvery grey, vertical-striped parore. They are like wild sheep migrating to mountain pastures to graze. Normally they live among the coastal seaweed forests, but with the tide they enter estuaries and swim upstream to graze the stubble of algal fuzz from the submerged mangrove trunks and branches. At the base of a tree I see a broad tail swirl. With a sideswipe of its head a parore rakes off a mouthful of greenery with its three-pronged teeth. For some reason a few of the school choose to stay overnight in the river when all the others depart.

If I remain very still, schools of grey mullet come zooming out of the blue like silver arrows and I can watch them feeding on the mud, dipping down to gulp a mouthful and blowing out a shower of particles from their gills, having filtered out any nutrients.

All the way down the river I note the location of the most interesting stretches of overhanging bank, sunken logs and other special features, building up a mental map of the streambed so that I will be able to find my way in the darkness.

Nearing the sea I notice a pipi, bivalve mollusc, gliding over the sand-rippled streambed. Something odd about its attitude: without tumbling or turning it slides downstream, its long axis in line with the current. Then I notice a long skein of mucus extending ahead of it, extruding from its leading edge. I glance around. Squadrons of pipi are moving down the estuary in the same manner, each tugged along by a release of mucus that is acting like a parachute. Suddenly I recall Dr Roger Grace's observation: pipi use this parachute technique to hitch a ride on the currents – an amazingly efficient strategy for a simple bivalve, paralleled by the aerial voyages of certain spiders that drift across oceans on gossamer threads.

MANGROVE NIGHTS

Now I stagger down to a deep stretch of river laden with diving light, photo gear and scuba. High excitement! This is like Friday night downtown: life forms are darting or lurking everywhere, and each sweep of my torch reveals a startling tableau. Flounder are hunting crabs. Waves ripple along their peripheral fins and tails undulate or swerve. Normally their chameleon-style camouflage renders them invisible on the sandy mud, but my light throws their bodies into relief. I almost choke with laughter. Salvador Dalí! An eel meanders towards me with a crab upside down in its jaws. The legs stick out like the painter's waxed moustache.

Fleeing from my light are clouds of small, wine-red fish that look strangely familiar. Nocturnal bigeyes from the ocean, up in the mangroves to feed on plankton. With their huge eyes they can select tiny prey by starlight.

In a cavernous overhang, transfixed by my light, is a resting parore, its body has darkened, stripes faded, a series of whitish blotches along its back. This is its disruptive night attire. By moonlight it breaks up the body outline, providing

camouflage. Then I see a row of eels at the back of the cave staring out like passengers on a bus. Why are they lurking in this dark recess? Is this some sort of predator trap?

Out on the riverbed again a few drowsy blennies huddle close to shelter, no longer dancing over the bottom or posturing defiantly. For the most part they have found total concealment. Startled by my lights, a grey mullet jets about colliding with obstacles, confused and disoriented. During the night, school fish like this just hover individually mid-water. Their schooling behaviour no longer offers any protection.

To my surprise a large kelpfish rests among rocks by the bridge. This reef dweller hunts only for crabs that stir at dawn and dusk. Perhaps at twilight it prowled upriver and chose to settle up here for the night.

Here and there I pass drowsy babies: juvenile snapper, chocolate striped, hovering in depressions; miniature triggerfish, wrasses, goatfish and trevally, all fins erect to maximise size. Mangroves are important fish nurseries, but this is most apparent at night. Paradoxically this is also the time when predators are busiest. Perhaps all the crabs and other crustaceans keep them occupied.

ABOVE, TOP TO BOTTOM: At night, parore *Girella tricuspidata* with white-barred camouflage 'pyjamas'; a drowsy juvenile snapper *Pagrus auratus* in its mangrove nursery.

OPPOSITE, TOP TO BOTTOM: Shyest of all the mangrove birds is the banded rail *Gallirallus philippensis*; along the bank the white-faced heron *Egretta novaehollandiae* stalks on matchstick legs; a pied shag *Phalacrocorax varius varius* dries its wings after feeding below water.

BIRDS IN THE MANGROVES

From my platform up in the canopy I watch the birds. All predators, they are at the top of the food pyramid. They have few enemies. They control populations and they influence selection. Only the fittest of creatures can avoid their quick beaks.

Shyest of all the mangrove birds is the banded rail. One day at low tide I noticed semicircular ripples radiating from a low bank. With tele lens I pried. It was a rail. I learnt that each day this bird crept down through the trunks and tangled roots to a regular bathing spot at the water's edge. Repeatedly it dunked under with lightning speed and preened its feathers. It is said to be very vocal at dusk. I have yet to distinguish its voice.

Along the bank stalks a white-faced heron on matchstick legs, its long beak and sharp eye swivelling to and fro, intent on the water close to the bank. Then, near the edge, I notice a trail of bubbles and a flash of white. A pied shag surfaces, gulping down a small eel. I came to realise this was a cooperative hunting technique. Each bird scares fish towards the other. I have never seen herons hunting together; nor pied shags. But these two fish hunters have overlapping habits and a good basis for a symbiotic partnership. Later I saw a royal spoonbill, immigrant from Australia, walking up a shallow streambed, its head weaving in arcs as it raked the water for plankton with its bristle-lined beak – a bit like a humpback whale!

An individual shag claimed the stretch of river near my platform. This was his special feeding territory. Each day he would return from the coastal pohutukawa tree where his diving colony roosts. He has no rival on the river. One dead branch is his wing-drying perch. In order to dive efficiently shags dispense with the oily, air-filled feathers of other seabirds and find it necessary to preen and dry after each underwater hunting session.

Kingfishers are perched at intervals above the river, ready to flash down and scoop up a crab, blenny or sprat without ever landing. Their sharp staccato cries, in sets of three or four, establish hunting territories, just as the morepork demarcates its night zone after dark and swoops up the tree-tunnel river.

Fantails flutter around me, snapping at insects with audible clicks of the beak. Sometimes a fantail's claw manages to seize a large moth on the wing

and it perches one-legged to eat its prey from the other claw as would a parrot. Scrambling up branches and hovering amidst foliage are tiny ruby-eyed grey warblers searching for insects. From time to time one hovers in mid-air exactly like a hummingbird, inspecting a bunch of leaves. Elegant silvereyes dance about after other grubs and insects. I reflect on the service these birds provide the trees. Without them all the leaf chewers and sapsuckers and wood-borers could reach plague populations.

Further up the estuary in the salt marsh and raupo swamp are the haunts of the bittern, fernbird, spotless crake and pukeko. For all these birds the mangroves are just an extension of the native forest, a happy hunting ground where reef fish and bush birds mingle sometimes only centimetres apart.

VIDEO DIVES

In high summer, a warm mid-January day, in part because I suspected an upriver swim might avoid panicking the larger fishes from fleeing towards the sea, I set out an hour before full tide from the mangroves just below the Matapouri bridge. I wore scuba so I could concentrate on my subjects. I also had a video camera.

Like timid sheep large herds of parore race upstream as I approach the shadow of the bridge. Its abutments constrict and hasten the current. The bottom is flushed clean, a firm bed of gravelly sand with scattered rubble and small rocks. Shelter for a megalopolis of triplefin blennies. No longer furtive and suspicious of my camera as they can be in the open sea, these bottom jumpers keep crowding around my lens. Perhaps they suffer very little predation here.

Miniature spotties, a species of wrasse, flit about, pecking at the encrusted rocks, and crabs scuttle for cover. Suddenly a flounder spurts across my path. A flounder in broad daylight! To my surprise it tolerates close attention. Moving constantly it is bottom feeding, dipping down every few seconds to gulp the sediments, squirting the detritus from its upper gill. I am so close I can see the exhaust particles that collect in drifts on its back. Just behind the gill the pectoral fin is erect like a small mast, twisting and bending. It acts as a forward rudder while the whole fish undulates rapidly from stem to stern. Obviously the flat tail would be of little use for steering.

Further on the river deepens and narrows with a firm sandbank on one side. On the sand crest lies another flounder. Just its gill cover pulsing. Above its protruding peacock-tail-blue eyes a tiny scrap of silver hovers and rotates: a baby trevally. And nothing induces it to flee. I can approach within centimetres and

try several angles. Both fish remain. Related to the tropical pilot fish that ride the pressure waves on the noses of sharks, juvenile trevally often adopt a larger fish for companionship.

Now I am approaching a major bend in the river, a point where it converges with a smaller stream. A storm flood has swept a mat of roots into a vertical stance like a hedge. Alongside it a clean sandy hollow has been scooped out; a dense sward of eel grass is adjacent. For some reason this location is one of the 'hottest' spots in the whole river. And today the activity is frenetic. A rope of yellow-eyed mullet winds past me to gather at the surface and form a frenzied circle of plankton snatchers. Others are angled like knife blades, heads buried in the eel grass, bodies flashing as they nip off the tiny plants and animals clinging to the leaves. But they have competition. A herd of young parore is similarly engaged. Every day they feed here yet the eel grass bed is totally intact. Then I make a discovery: they are grazing the algal fuzz! The fine growth of epiphytic plants they harvest from the blades will regrow as fast as my beard – and feeding time is limited by tide and daylight.

Among the thicket of roots a large eel has insinuated its snake-like body. Further along, the crest of barnacle-encrusted roots opens out like the tines of deer antlers. Clouds of very young parore weave among them, grazing the algal

Juvenile trevally *Caranx georgianus* 'nursed' by a sand flounder.

fuzz. Miniature wrasses pectoral paddle hither and yon, nipping at barnacle cirri: juvenile spotties with twin black beauty spots on dorsal and anal fins that disappear with age.

I find a newly killed blenny, its head severed. Whelks feast on its flanks. A group of blennies are poised in the vicinity watching but never intervening. Such forbearance by fish seems odd. These blennies are vigorous scavengers. But experimental studies of blenny learning capacities show unexpected brain development. Do blennies have such social bonds they shun to eat their kin?

Up on the mud bank I watch a large mauve rock crab manoeuvring like a Sherman tank amidst the necklace weed. Beside it a pinhead-size anemone on a cockleshell expands like a flower. Then the blenny pack advances, boldly peering at the crab within claw range. The brat pack! Experimentally a teenage blenny makes a swipe at the crab. Another attacks. Massive armour wins the day and the mauve crab continues its journey unperturbed. But further on a smaller crab has not been so lucky. A legless carapace is locked in the jaws of a big blenny. The rest of the gang are focused on the hunt, hopeful of snatching the prize. A few try. The big one gulps and chomps several times and the crab is engulfed. The reward of age!

FROM RIVER MOUTH TO BRIDGE

From upstream my thoughts turned to where the river meets the open sea. I decided to follow the sequence of living conditions from the river entrance up to the pool below the bridge to see how things cope between the jaws of hell and paradise. The break at Matapouri river mouth is a popular learning ground for young surfers. In big storms waves pound into the semicircular bay and disperse their energy here as they collide with the rain-swollen river. What could live in such a zone?

Near the top of a rising tide I venture out with my heavy camera on a calm summer's day. With the scouring action of shifting sand and low-tide exposure to the fierce sun, life is tough here. Dwarf mussels and a carpet of surf barnacles smother the upper surfaces of the rocks. On some vertical faces a white, curling mass of tubeworms thrive, delicate blue gills that vanish at my presence into their strong calcareous tunnels. A tiny, lone goatfish spurts from my camera. Small hermit crabs scuttle over craggy rock faces where lined whelks scavenge. Above them on sunny crests fast-growing tufts and lacy sheets of brown kelp

TOP: Wade beneath river downstream from Matapouri bridge with a video camera to film gobies.

LEFT: On the broad, flat riverbed, the twin siphon holes of cockles *Chione stutchburyi* dot the bottom.

ripple. Any seaweeds surviving here would have rapid growth and a limited lifespan. I follow the tide upstream over oyster-sharp reefs and shell-littered sand where blennies leap and dive for cover. A thin scattering of emerald sea lettuce.

A clump of fat oysters have colonised a branch that leans down into the river from the bush-clad clay bank. Even a young kauri tree stands with its roots in salt water. Then a whale shape looms into view. The massive trunk of a pohutukawa tree that chooses to reach out horizontally into the river, providing a diving platform for children. In its lee I find a smoke puff of minute larval fishes like silvery seeds racing about. Ghost shrimps tread delicately over air bubbles trapped in the bark.

Just beyond the first bend in the river I meet the mangrove advance party – a fringe of small bushes that expand in size every few metres along the bank. Out on the broad, flat riverbed the twin siphon holes of cockles dot the bottom. Feasting on them are whelks clustered in mounds. A few cushion stars glide over rock patches. Small colonies of sky-blue tubeworms encrust lumps of rock. Fringing the bridge pool near scattered mangrove bushes low reefs support eel-grass beds where parore graze. Among the blades a ghost shrimp explores. In the eel grass I find the tiniest of tiny spotties in dense numbers, so well camouflaged it would be very easy to overlook their presence, but fascinating to watch centimetres from my mask as they go about their browsing business in this mangrove haven of calm.

THE VALUE OF MANGROVES

An early New Zealand novelist wrote of mangroves with loathing: 'I never saw one that looked as if it possessed a decent conscience. Growing always in shallow, stagnant water, filthy black mud or rank grass, gnarled, twisted and stunted and half bare of foliage, they seem like crowds of withered, trodden down old criminals, condemned to the punishment of everlasting life.'

Such an attitude to mangroves has been shared by generations of New Zealanders who live in the mangrove zone. Northland and the Hauraki Gulf, including harbours like the Kaipara, Waitemata and Manukau, have a vast extent of deeply indented coastline. Almost every bay, estuary and backwater in this area is fringed with mangrove trees. They reach their greatest sizes around Whangarei and further north, where they line tidal rivers in a dense green canopy up to 8 m in height. Further south they are much smaller and stunted.

Mangroves are scarcely more gnarled and twisted than the beautiful pohutukawa. The fact that their environment may become foul and pestiferous is largely the fault of humans who restrict the tidal flow around them. After years of exploring them above and below water, by day and by night, I have come to see them not as 'trodden down old criminals' but as a vital element in our coastal ecology. For too long it has been the custom to regard mangroves as wasteland, fit only for the dumping of refuse or for reclamation.

When claims are made about mangroves spreading, there is no regard given to the large areas now farmed. For the sake of a few hectares of swampy grazing land we lose the cradle of the sea: a huge interrelated network of living systems which scientists as yet do not fully understand. Like the radiator of a car it offers a maximum of exposed surfaces to the sea. You will have noticed how swiftly any new jetty pile, boat hull or drifting object is colonised by marine animals and plants. Settling surfaces are at a premium in the sea. A shipwreck can transform an area of barren seabed into a thriving community, just as an oyster or mussel farm is an artificial settling surface enabling shellfish to establish themselves in the self-fertilising sea. In this light it may be appreciated how serious it is for our coastal and marine environments if we persist in 'tidying up' our scruffy harbours and estuaries with neat paddocks, highways and rubbish dumps.

The oceans cover two-thirds of the planet's surface, but their most prolific zone is the shelves of the continents, stretching from the seashore down to 200 m. This is the greatest domain of life on Earth. Unfortunately it is also the area of the sea most subject to the consequences of our pollution and interference. We have to break the habit of treating the open ocean as an infinite resource for a hungry world. For that we need to keep intact every tide pool, every mangrove, every piece of shoreline.

Sand and gravel beaches provide an insecure and shifting environment for a reduced range of life forms. Those adapted to a burrowing existence form very dense populations in such areas – pipi and other bivalves, tubeworms and swimming crabs. Further up the shore the sand dunes provide a very fragile habitat in which the least disturbance of plant cover can cause erosion.

The richest marine worlds are the most sheltered parts of our coast. Mud is the basis of it all: in thick, black mud thrive millions of bacteria, which readily convert decaying matter into nutrients, a process on land well known to soil technologists. The difference between clay and fertile soil is the presence of countless useful bacteria, which convert animal manure and leaf mould into plant food.

In a forest when a leaf falls the solar energy store it represents usually stays within the forest as leaf litter, but in mangroves each fallen leaf releases its nutrients to the tides and beyond. In all the world's oceans, mud and ooze swarm with bacteria. From the upper waters a rain of dead animal and plant life is attacked by them and broken down into nutrient salts and minerals. Over the continental shelves upwelling currents bring these recycled organic chemicals back into contact with sunlight in the upper layers of the sea. Here microscopic plant plankton feed on them, generating a rich algal soup which nourishes the whole marine food chain from animal plankton to sardines and great whales. Such is the importance of algal plankton that it is now believed they are the main source of oxygen in our planet's atmosphere and chiefly responsible for its renewal.

For algal plankton the shallow, warm expanses of harbour and estuary are a hidden pasture of the richest resources, for here nutrients flow from the land and from the sea. Twice a day, into every sheltered place on the New Zealand coast, a tidal top-dressing device glides in under moon power. Along estuarine shores where wave violence is minimal; over harbour flats where exposure to the sun is seldom fierce; amidst grasses, reeds and leafy stalks it flows, bringing fresh food and oxygen to all the crabs and snails, worms and barnacles, shrimps and oysters, which duly grow fatter, multiply and release their larvae to the sea. As each tide recedes, fishes gorge themselves on the rich outpourings, wading birds begin to stalk over the refurbished mudflats and on sandbanks cockles and tuatua close their filter intakes until the next tide.

Mangrove glades play an important role. On every fine filament, twig and rootlet, silt is deposited. All rivers carry silt from the surrounding hillsides. Usually it flows out to sea. If a river is slowed down with serpentine windings and increased vegetation, the silt will be retained. This the mangrove does extremely well, being the only plant able to withstand the harsh conditions at the meeting place of land and sea. Its roots and branches stabilise mud banks. Over what would otherwise be an intertidal expanse of smelly mud they extend very gradually to reclaim the land.

If we alter the mouth of a river, removing mangroves and straightening things up, we must realise that in so doing we increase the flow of the river right back to its headwaters. Without the network of mangroves at its mouth, and the meandering turns, the swift flowing waters bear an increasing volume of silt out to sea, causing erosion in the hinterland and affecting our fishing grounds. Nature loves the wild disorder of the mangrove forest. And so do an increasing number of those who enjoy outdoor pursuits.

The vast settling surface offered by the mangroves' contorted roots and trunks, bathed twice daily in a nutrient tide of salt water, provides a nursery for the pelagic larvae of many creatures from the adjacent sea. Each outgoing tide releases rich swarms of tiny crustacean larvae and other living material back into the sea where it helps support our inshore fish populations. Mussels and oyster spat have a tremendous mortality rate in the open sea.

It is the huge settling surface of our mangrove forests that enables sufficient

nutrients to be released for an adequate survival ratio. Sea plants and animals find a shoreline habitat very difficult to colonise because of the sharp contrasts in temperature, violent water movements and the drying out effect of tidal changes. Mangroves help them in that their roots stabilise these fluctuations. At low tide, water is held around them and there is shade and deep mud to reduce the extremes. With the rising tide the roots diminish the pressure of the current and quieten the underwater world during its main feeding period. Truly, the cradle of the sea.

TOP: Mangrove reclamation dredge pipe in Whangarei. For the sake of a few hectares of swampy land we lose the richest of marine nurseries.

BOTTOM: Mangroves stabilise the fluctuating movement of the sea.

Riverlands rata *Metrosideros robusta* and kanuka *Kunzea ericoides* both bloom at December's end.

CHAPTER 6:

OUR NATIVE PLANTS

While familiarising ourselves with the coastal trees and shrubs of Whale Bay Reserve, at the same time we were learning to recognise all those in our own piece of forest, some five km from the coast but running out to the tidal Ngunguru River. Over many years Jan and I had photographed the marine life on this coast. Now, in this forest exploration phase, equipped with digital cameras and guided by a checklist of plants recorded for our ecological region and aided by local experts, we set out to locate each species and trace its annual cycle.

Gradually our want list induced us to explore further afield – the Hugh Crawford Memorial Reserve, Brattys Bush Scenic Reserve, the Soda Spring Reserve and other Tutukaka Coast forest remnants – until we had completed our digital database. (In time, we would extend it to the whole of Northland.) Besides trees and shrubs, it includes vines, ferns, orchids, grasses, herbs, fungi and lichens and their kin. By this stage we had become painfully aware there were certain plants on the Tutukaka Coast that are rare or restricted to a few locations.

Karo *Pittosporum crassifolium* flower.

OPPOSITE, TOP TO BOTTOM: Ancient neinei or grass tree *Dracophyllum latifolium;* purple berries on rare koru *Colensoa physaloides* (synonym *Pratia*); hinau *Elaeocarpus dentatus* flowers.

TUTUKAKA FOREST TREES

At Woolleys Bay one spring day (mid-October) I sighted what I thought was an amazingly early pohutukawa in bloom. Out on a branch a host of bees were feasting on vivid red starburst blossoms. This I discovered was a stunning crimson rata vine (*Metrosideros carminea*). I know of only two others in this district. Why do plants become so rare?

On the slopes of Whale Bay in late October, we discovered hybrid lancewoods and, where the hill meets the flats, dozens of healthy bushes of locally rare New Zealand gloxinia (*Rhabdothamnus solandri*). Along the main track in early November the fragrant native jasmine has pendant blooms that develop into long black seed pods like skinny beans.

At Matapouri on a steep slope we found a few tree tutu bushes, so abundant in forests further north. And on the flat below, a strange puriri with pure white flowers rather than the usual pink. On a rock outcrop near the beach, coastal mahoe (*Melicytus novae-zelandiae*), another rarity in this area. Around the cliffs were karo with their dusk-scented deep-red flowers; on a hill ridge was an aged ngaio, leaves dotted with tiny oil glands; and in forest behind the estuary a solitary hard beech (*Nothofagus truncata*), fallen out of a landslip. In two beach areas we found the endangered prostrate fuchsia (*Fuchsia procumbens*), with exquisite cornflower blue and golden flowers. Beside the road south a small group of rare tree daisies (*Olearia solandri*), which bear masses of minute starry white flowers in mid-April.

At the Soda Spring Reserve, after having a refreshing soda drink from the bubbling forest-edge spring, we found nearby a massive puka vine (*Griselinia lucida*) writhing up an ancient totara, along with several big rata. Their hosts are a sign of old-growth trees. Possums delight in feasting on the lush, broad leaves of puka, and such mighty vines with their distinctive ribbed stems are becoming all too rare a sight.

On Tutukaka Lighthouse Reserve, close to the lighthouse itself, is a perfectly spherical, wind-compacted haekaro (*Pittosporum umbellatum*), along with several other rare shrubs. The Bowden property just to the north has a patch of rare king ferns, poataniwha shrubs and a very rare coastal maire tree.

Around the shores of Tutukaka Harbour we were

staggered to find a group of grass trees (*Dracophyllum*) on a forgotten ridge; a mountain maire above the marina; snowberry shrubs on terraces above the car park; and in a sequestered corner, a rare colony of *Pratia*, a lobelia-like native, with big soft leaves and blue-purple flowers and berries.

In a small DOC reserve just behind Tutukaka stands the giant kauri Tane Moana. Jan and I measured it carefully. A metre above ground level its girth is 11.16 m. It would be over a thousand years old. While not as tall, its girth is as great as some of the famous trees in Waipoua Forest.

In Ngunguru Reserve behind the public library the forks of a few pohutukawa trees support epiphytic shrubs of *Pittosporum cornifolium*, with their attractive scarlet seed capsules.

Up the river we have found another ngaio tree, brilliant scarlet rata and several grass trees. A mountain maire and haekaro both extend from the bush over salt water. Upstream near the Scow Landing, where fresh water begins, we found a huge titoki on the riverbank and a grove of sturdy ribbonwoods (*Plagianthus regius*), out in a paddock. Scattered through the area we know of a few hinau, pukatea, Kirk's daisy, tawa, miro, toro, matai, kawaka, akeake, tarata, tawapou and kohuhu; all sporadic or rare. The whau tree, lightest wood in the world and once an attractive part of this coastal forest, has now vanished from the wild here, but one lives on the saddle to Pebbly Beach.

At odd secluded places along Ngunguru estuary we have discovered native irises and rare colonies of mairehau bushes; their refreshingly aromatic leaves, stunning white flowers and red berries make me wonder why such a lovely native plant should become so scarce and forgotten.

WHY SO RARE?

I am deeply alarmed at the number of New Zealand plants we have found that are marginal or threatened. What are the causes of this serious loss of biodiversity? A great many of our New Zealand trees are berry producers and are dioecious (having separate male and female forms). Lack of birds to distribute them and gender isolation would be major factors. More than a century ago extensive forest clearance and burn-offs hit many parts of New Zealand hard, particularly the burnt-off kauri extraction zones. Many trees, now rare, struggled to regenerate. Another factor would be a lack of pollinators: birds, geckos and nectivorous bats. Scarcity of available habitat would be another. People developing lifestyle blocks and wishing to revegetate could have a major positive effect here if they were aware of the appropriate plants and their habitat needs. The current trend towards vigorous pest control on the Tutukaka Coast will reap huge benefits for all the pollinators, seed distributors and plants.

One of my green neighbours, to his horror, almost cut down the biggest,

SUBTROPICAL DIVERSITY

Northland forests are especially diverse because they support species with subtropical origins as well as temperate climate trees. The Tutukaka Coast is part of an ecological district extending north to the Bay of Islands and south almost to Whangarei Heads, in which forest types vary in a rich mosaic of patches. There are steep slopes and ridges where tanekaha, totara and kanuka are dominant. Elsewhere, kauri, towai and taraire. Puriri and taraire may prevail in a gully. There are slopes that are mainly towai mixed with rewarewa and puriri. A ridge may have a predominant stand of kauri; another, rimu. In some areas, like the view from my window, there is a nice mixture of kahikatea, rimu, kauri, totara, karaka, puriri, kohekohe, rewarewa, kanuka, nikau, cabbage trees and tree ferns. Similar locations are sprinkled with northern rata, and the occasional kawaka, tawa, pukatea, matai, miro or lancewood.

Some trees, like rimu and kahikatea, have regular conical forms. Others, such as taraire, kohekohe and kauri, have a sprawling canopy. This creates variations in light access that offer opportunities for smaller trees as a second canopy or understorey. Depending on situation this may include pigeonwood, tree fuchsia, houhere (lacebark), kowhai, red matipo (mapou), five-finger, pate, hangehange, mamangi and kanuka. On forest margins and in light wells there is an edge community that may include kumarahou, mingimingi, hebe (koromiko), kawakawa, poroporo, korokia, rangiora, wharangi, tree daisy, manuka, taupata, karamu and other small coprosmas, the bush daphne (*Alseuosmia*), the New Zealand gloxinia, cabbage trees of two species and flax. Then, there are all the vines and other plants . . .

OPPOSITE, TOP TO BOTTOM: Multi-toned kauri leaves *Agathis australis*, new growth and old; karamu *Coprosma robusta* berries; kawakawa *Macropiper excelsum* fruit; kohekohe *Dysoxylum spectabile* fruit splits.

LEFT, TOP TO BOTTOM: *Veronica bollonsii*, a coastal shrub; kanuka *Kunzea ericoides;* kohekohe *spectabile* flower; pigeonwood *Hedycarya arborea.*

most umbrageous putaputaweta tree I have ever seen, thinking it was a cursed privet. Another, a great lover of plants, was halfway through demolishing a magnificent native passionfruit vine believing it was an exotic wind-spread moth plant. We New Zealanders are often unfamiliar with many of our finest plants and risk losing them forever as our forests become bird silent, our landscape exotic. Areas of New Zealand like the Tutukaka Coast have an immense future as refugia: oases of wildlife belonging to the ancient world of Gondwana. Most of New Zealand's trees grow in the north – and quite a few extras!

FAVOURITE TREES OF THE TUTUKAKA COAST

TANE MOANA

Jan and I have made pilgrimages to the major named kauri trees of Northland; we documented them with our cameras and wrote stories on each. We see these as living cathedrals; pyramids of the plant kingdom; blue whales of the forest. In a small DOC reserve just behind Tutukaka, recently accessible via the new Te Araroa pathway, stands a massive kauri that was almost forgotten. While not as tall, its 11.16-m girth places it among the great kauri trees of the north.

Tane Moana was poorly known until ritually named on 29 June 2008. Some 200 people of all ages walked out to the giant kauri that midwinter Sunday. Maori elder Sonny Wellington performed a powerful ritual, and a poignant waiata was sung. One middle-aged man present, Conrad Gray, was among our group that visited the tree at Christmas back in 1972 when we formed a circle of 24 linked arms around the immense trunk, vowing to protect the tree.

When I spoke, at Sonny's invitation, I told the story of the tree as I knew it, back to the time before any humans walked these shores when a seed like a cornflake blew to this spot and a vast tree arose that withstood centuries of cyclones and the felling of all its companions. I could not imagine 200 people walking to a tree back in 1978 when I first saw Tane Moana. Today there is a proper walkway around it, built by DOC. For a bronze plaque at its base, with Sonny, I wrote:

> One tree doesn't make a forest
> but Tane Moana proclaims our loss
> and determines us to preserve.
> Humbled by its might, sprung from a tiny winged seed,
> we reflect … *Kia whakaiti ia koe.*

OPPOSITE, TOP TO BOTTOM: Reggie Hole, our neighbor, measuring Tane Moana, February 2008; a protective group around the as yet unnamed Tane Moana, Christmas, 1972.

The kauri is a cone-bearing tree with small brown male cones like catkins and large emerald female cones on the same tree. Its winged seeds spread with the wind. The young tree, or ricker, is a typical pyramidical form, but as it grows its shape changes radically: it sheds the lower branches from its trunk and rises some 50 m into a tall, smooth column topped with a massive spreading crown. Impressive stands of kauri dominate ridges in the Crawford and Brattys Bush Reserves. There are small groups at secluded spots around the shores of Tutukaka Harbour.

KAHIKATEA

Next to the kauri I have a special regard for the kahikatea (*Dacrycarpus dacrydioides*). It is our tallest-growing tree and can soar to 60 m. It's another cone bearer, but, as with related totara and rimu trees, there are separate male and female trees, and birds spread their seeds. Each tiny blue female kahikatea cone has a golden fleshy receptacle attached. In autumn during good seed years the female cones ripen, giving the tree a golden glow, while male trees appear blue. Native pigeons and tui go mad feasting on these tiny nourishing spheres. Maori once went to immense effort to harvest them, too, and early settlers made kahikatea jam.

PURIRI

Just inside the stile at Whale Bay Reserve and near a picnic table a puriri (*Vitex lucens*) tree grows on the windy ridge. Through a hobbit hole at its base I can vanish inside! The entire tree is hollow: the trunk and immense boughs have all been cored out by the grub of the puriri moth. But the tough old tree has withstood every storm. Puriri trees are beloved by birds, too, because they offer their cherry-red berries all year around.

REWAREWA

While exploring the springtime forest with a young German visitor I noticed she had picked up the striking bloom of a rewarewa (*Knightia excelsa*) tree and treasured it all day long. That Fräulein made me aware of what an especially intricate and particular beauty the rewarewa has with its velvety red, gold-flecked flower clusters, each about 10-cm long, an array of tightly coiled, spiralling individual flowers. A straight, slender tree, the rewarewa grows fast and can reach 30 m in height. It even thrives on dry, exposed ridges such as Whangarei Heads. The flower develops pods resembling tiny canoes that burst open to release winged seeds on the wind.

MAIREHAU

Plants once regarded with the greatest distinction in early Maori times can slip away to near loss in the world of today. This is why I champion the mairehau (*Leonema nudum*) shrub. Related to the citrus family and resembling a manuka or mingimingi, this plant has leaves so fragrant they are almost intoxicating. Maori maidens used to anoint their bodies with an oil perfumed with it prior to lovemaking. Jan and I enjoy exploring the hardest to reach zones along the Ngunguru River. In places too shady and steep for fire, cattle or goats, we find relicts of mairehau. Imagine our delight when Jan chanced a really big one as she swung down a steep, newly made rat-baiting track on our land and stretched for a handhold. It was in bloom! Mairehau produce sprays of pristine white flowers in spring and small red berries, much loved by birds, in autumn.

KOWHAI

On a golden September morning Jan and I set out to survey all the kowhai (*Sophora chathamica*) tree assets of the Tutukaka Coast. Above Whangaumu Bay, with the twin peaks of Whangarei Heads in the distance, golden patches were clinging valiantly to the cliff edge after recent assault by the most powerful winds this coast has known in decades.

Then, around the sheltered shores of Tutukaka Harbour, we admired the gilded dome of the volcano plug Philip Island out in the middle of the bay, glowing in the late light like one of those cupolas in Paris. Above Church Bay adjacent, the headland gleamed with yellow kowhai groves as nectar-feeding tui flitted from island to promontory. Kowharewa Bay extended two capes spangled in gold.

Kowhai are one of New Zealand's few deciduous trees. As legumes they are nitrogen fixers. This gives them a nutrient advantage on harsh cliff-face habitats. For much of the year they are hardly noticeable among the diverse array of cliff foliage; even less so in winter when they become bare of leaves. This heightens the impact of their August–September bloom time when their bare branches are adorned with clusters of golden trumpets. Kowhai *want* to be seen: attracting pollinators with sweet bribery is their strategy. Each bright tree is a beacon to the tui community.

So this calm and shining morning we decided to do a thorough stocktake of all the kowhai groves along our headland coast. The cliffs above Woolleys Bay were the first major display. East-facing hills in Matapouri had a smattering of gold, but the headland groves facing west were fading to pale green as foliage replaced flowers that were now preparing to develop into long drooping seed pods like knobbly peas. Both here and at Ngunguru, above the river entrance, we noted that west-facing groves tended to be the earliest to bloom and first to

fade. Staggered flowering times extend the nectar flow and avoid overloading the feeding birds.

Under a kowhai grove at Whale Bay, tui swooped low overhead. A colony of squawking shags joined the sweet voices of the forest birds like harsh punk rockers as they vandalised a skeletal pohutukawa with their abundant droppings. A chirruping flock of silvereyes swarmed over a dense spray of kowhai flowers, hanging upside down or tumbling about as small, sharp beaks probed for sugary drops left by the tui. A pair of grey warblers chased each other in a furious display of aerobatics. Then one bird broke off and flew in a beeline to a manuka bush on the cliff brink. In a fork at its centre I distinguished a small, vertical oval mass shaking and vibrating. Somewhere in that down-lined grass cocoon there would be a circle of hungry beaks. The warblers would be eager to raise a batch of chicks before the shining cuckoos arrived from their winter habitat in the tropics to usurp the warbler nest, expel their young and replace them with alien eggs that the warblers would foster.

Among the flowers of the kowhai I observed, tender young leaves were beginning to develop like ferns, unfurling from the feathery tip into a double row of leaflets like small canoe paddles. In the summer months each pollinated flower would evolve into a long seed pod full of wheat-grain-sized 'peas' with a shiny yellow, steel-hard coating, so tough that plant propagators soak them in boiling water to trigger them to sprout. I wondered how such seeds might be distributed from clifftop to water edge? They do not seem to be bird food. Along tide margins the seeds float and would travel widely by water, but how do they get to the hill summits? Perhaps fierce winter gales scatter the winged seed pods up the cliffs. Or is it by pod explosion?

KAKA BEAK

And then I remembered the lovely relative of the kowhai: its crimson cousin, the kaka beak (*Clianthus puniceus*), a shrub now so endangered it survives only in a few small patches on the bony flanks of offshore islands. Fortunately, in the hands of skilled gardeners, kaka beak can thrive. Its bloom is as exquisite a rival to the jeweller's art as any neck pendant: an elegant scimitar of bright scarlet set in an emerald green, tooth-edged cup. In one garden we saw a very rare white kaka beak.

TREE FUCHSIA

Another of my favourite deciduous natives is the tree fuchsia or kotukutuku (*Fuchsia excorticata*), with its papery bark. It lives along stream edges and is extremely vulnerable to the ravages of possums. Like kowhai, tree fuchsia serves

TOP, LEFT TO RIGHT: Kahikatea *Dacrycarpus dacrydioides*, orange fruits and blue seed cones; rewarewa *Knightia excelsa* flowers form a cluster like tiny canoes.

CENTRE, LEFT TO RIGHT: Mairehau *Leonema nudum* flower; kowhai gold *Sophora chathamica*.

BOTTOM, LEFT TO RIGHT: Puriri *Vitex lucens* bloom and berries; kaka beak *Clianthus puniceus*, a relative of kowhai; tree fuchsia *Fuchsia excorticata* flower.

up its treasures in springtime: flowers of such intense colour and delicacy they take my breath away. The pollen at their golden centres is deep blue; the only blue pollen in the world. In autumn its sweet konini berries ripen for the birds, and for those aware of their pleasant taste.

VINES OF THE TUTUKAKA COAST

CLEMATIS

At peak abundance on the Tutukaka Lighthouse promontory clematis is one of the first manifestations of springtime on this coast. Its stunning blooms garland the canopy of manuka and kawakawa bushes with a bridal finery of virginal white, lemon- and pink-centred stars; rococo arrays of flower panicles, lobular dark green leaves, swelling buds and twining stalks. Each male clematis flower is an elegant diadem fit for an elvish king. In advanced female flowers this opens into a whorl of windwaft, gossamer-winged seeds that Jan and I like to call 'the fuzzies'. Bees tumble and hover in orgies. *Clematis paniculata* vines on this headland are said to be denser and more accessible than anywhere in New Zealand.

On the slopes of Whale Bay our plant list expanded when, in late October, we discovered the lemon-scented clematis (*C. foetida*); a new record for the area. Not far away, along the borderline fence in the Tahere Reserve, and near an impressive waterfall, we found a third species, the dainty little green clematis (*C. cunninghamii*).

TOP, LEFT TO RIGHT: Puawhananga vine *Clematis paniculata*; scarlet rata vine *Metrosideros fulgens*.

CENTRE, LEFT TO RIGHT: New Zealand jasmine vine *Parsonsia heterophylla;* broad-leafed bush lawyer vine *Rubus australis*.

BOTTOM, LEFT TO RIGHT: Kiekie or pandanus *Freycinetia banksii*; New Zealand passionfruit vine *Passiflora tetrandra*.

RATA

This district has four kinds of rata vine. Most common is the white rata (*Metrosideros perforata*), with its tiny round leaves pocked on the underside with small pits. It blooms at Christmas with masses of bee-loved white flowers. Less obvious but quite abundant is the scarlet rata (*M. fulgens*), with bigger, oval leaves. Because it only flowers in full sunlight we often miss its vivid autumn blooms unless we can look down on the forest canopy from a hillside.

Just west of the Crawford Memorial Reserve on the Tutukaka Highway, access to the Waipaepae Reserve has to be negotiated through private property. Along an eastern fence line we homed in on one of New Zealand's most remarkable trees: a nikau palm garlanded just beneath its crown with a wickerwork sphere of interwoven rata vine, sprigged with diffuse pink flowers

THE FIGHT FOR LIGHT

Rata vines and bush lawyer, clematis, pandanus (kiekie) and supplejack are all vines or lianas. In a rainforest with so many plants competing for space, sunlight is at a premium. Getting up into the canopy and gaining enough light is a real battle. The vines short-circuit the problem. They don't waste time and energy producing a big trunk. Their tendrils just run up others and spread. Some, like the rata and puka – and the strangler fig in the tropics – completely take over the host tree and replace it with a stout trunk.

Metrosideros diffusa rata vine, rare in our area.

and pale- green oval leaves. This was a pink rata (*M. diffusa*) in full bloom. During our exploration we found nine of these rare vines: three pink, four white and two uncertain.

On another spring day (mid-October) at Woolleys Bay I sighted what I thought was an amazingly early pohutukawa in bloom. Out on a branch a host of bees feasted on vivid red starburst blossoms. This I discovered was an *M. carminea* rata vine. I know of only two others in this area. Not far up the coast, near Whananaki, we have seen a huge *M. albiflora*, a large white-flowered rata which has larger, pure white flowers. This species is common over on our west coast.

JASMINE

Along the path down to Whale Bay are three vines of New Zealand jasmine (*Parsonsia capsularis*), the only ones we have seen in the district. It has clusters of nose-worthy, fragrant white flowers like tiny bells that develop into long stringy, bean-like seed pods. As the vine spirals around a host trunk, often a kanuka, it binds so tightly the tree is left with a deeply grooved spiral all the way up. Perhaps it would make an impressive walking stick.

BUSH LAWYER

There are two kinds of bush lawyer on the margins of our forests: the round-leafed lawyer (*Rubus australis*) and the long narrow-leafed one (*R. cissoides*). Related to the raspberry, they both produce clusters of white flowers in spring, very like those of the exotic fruit. Edible berries follow but vanish rapidly into the beaks of birds, and so are dispersed. The sharp spines on the underside of lawyer leaves are not for defence but enable the climbing vine to hook onto support plants. So watch out for a scratched face from either of these lawyers.

PANDANUS VINE

'Maori bananas' is another name for the mushy golden fruit of our native pandanus vine (*Freycinetia banksii*), better known as kiekie. Few people ever see the fruits these days because rats reach them first. In the forest kiekie vines sprawl around the bases of trees until they take hold and climb. On tall nikau palms they look spectacular, festooning the trunk all the way to its crown with luxuriant flax-like leaves as if it were decorated for a festival.

PUKA

There is debate as to whether the puka (*Griselinia lucida*) is truly a vine because, while its strong, fluted cables ascend the trunks of old-growth trees, often in company with rata, and expand into lush, broad-leaved masses up in the crown, in time the puka will replace the host tree and become a tree in itself. Where possums are rife puka are very hard hit.

PASSIONFRUIT VINE

In his landscaped garden up Waiotoi Road, Bob Johnson showed us a huge vine that is rare in this district: the New Zealand passionfruit or kohia (*Passiflora tetrandra*). It was covered in profuse new growth that had been stimulated by his effort to cut it down, thinking it was a bad exotic. At the base masses of lush, slender green tendrils had burst forth and were heading skyward, clinging to the peeling bark of a big old totara. A dense curtain of glossy dark-green, willow-like leaves swayed in the wind. How could such a handsome native plant be vanishing, wondered Bob. We looked up the curtain of greenery to find masses of yellow flowers all the way up to the canopy. Exquisite miniature passionfruit flowers were sprouting from leaf stem junctions in a spiralling sequence. The vine was in full bloom: November. Bob's axe certainly got things going. But these were female flowers. Without a male in the vicinity they have never produced any fruit.

Each male flower has four orange-brown anthers. Although our native passionfruit produces inedible fruit, the small golden spheres adorning its green tresses make it the most desirable of decorative plants and a favourite of the birds. Despite every effort I have not yet been able to propagate it or purchase it from any nursery.

SUPPLEJACK

Sometimes in our forests I encounter such a dense entanglement of supplejack or kareao (*Ripogonum scandens*), it is utterly impenetrable. I have heard that the early New Zealand explorer William Colenso recorded finding a man's bones enmeshed

in a supplejack thicket. Hard to believe, but if he was so determined to pass through, perhaps being pursued . . . On the positive side, this flexible, jointed vine was a very useful plant to early Maori for fish-trap construction, as a source of water, and for its fast-growing (5 cm per day) edible asparagus-like tips. In autumn the bush glows with its vivid red berries.

PROSTRATE FUCHSIA

In two beach areas and up Ngunguru River near Cape Horn Cemetery, entangled among flax bushes and long grass, we have found the lovely prostrate fuchsia (*F. procumbens*), with its gem-like cornflower blue and golden flowers, exquisite enough to die for. It is increasingly rare because of grazing animals. Its large, sweet purple berries were once sought after by foraging Maori kids.

SAND BINDERS

On sandhills and places between the sea and forest, tangled masses of wire plant (*Muehlenbeckia complexa*) used to abound. It is still present but much diminished. Its billions of tiny green flowers turn into minute nuts on which native birds gorge. Our native copper butterfly depends on it utterly as host plant for its larvae. Wire plant binds the sand and provides a rich habitat for many invertebrates, which our birds also feast upon. More common inland, but evident by the Waiotoi River, is its relative with larger leaves, the tree-climbing *M. australis*, pohuehue, which can develop a very strong thick wooden stem.

Sprawling out over the hot sand at Sandy Bay is

TOP TO BOTTOM: Supplejack vine berries *Ripogonum scandens;* prostrate fuchsia *F. procumbens*; native sand-binding convolvulus vine *Calystegia soldanella.*

OPPOSITE, TOP TO BOTTOM: Silver tree fern *Cyathea dealbata*; kidney fern *Trichomanes reniforme;* young growth of pink fern *Doodia australis.*

our native convolvulus (*Calystegia soldanella*), with its mauve and white flowers: another valuable sand-binding plant.

FERNS ON THE TUTUKAKA COAST

Before there were any trees the ancient world was full of ferns. They are still with us. Some have underground stems or runners, which, as with bracken fern, were an important Maori food source. A few have edible fronds. Tree fern trunks formed the walls of dwellings or fences. Our mamaku or black tree ferns (*Cyathea medullaris*) are said to be the biggest ferns on earth. They are fast growing and can reach some six metres in height. Many on our coast died off in recent droughts. Pith from the upper trunk of mamuku was a starchy food for early Maori; it was steamed for two days to form a sago-like substance.

There are four species of tree fern in our district, each adapted to dryer or moister conditions, from the ponga or silver fern (*Cyathea dealbata*) on higher slopes to the wheki-ponga or golden tree fern (*Dicksonia fibrosa*), and the soft tree fern of stream edges: *Cyathea smithii*.

We find the most delicate of filmy ferns deep in the forest, often growing over logs. And out on the coast, small, leathery spleenwort ferns cling to rocks in the spray zone. My favourite filmy fern is the kidney fern (*Trichomanes reniforme*). In dry weather it curls up to almost nothing, only to expand into a diaphanous emerald circle when the rains come. You can almost see through kidney ferns because they are only one or two cell layers through. When they reproduce, the perimeter of the frond edge develops tiny, darker green nodules called sori,

which embellish the fern and make it even more impressive. Within these tiny cups spores develop, ripen and burst forth to reproduce the fern. Before plants developed seeds, which provide a useful food store for the young plants, microscopic spores were the predecessors.

Ferns clad our stream edges and carpet the forest floor. They can survive in low light levels and high humidity. Some ferns behave like vines and spiral up tree trunks, giving our rainforest its luxuriant texture. Many ferns are epiphytes and perch prettily on branches. To illustrate just how widespread and adaptive ferns are, the scented fern or pig fern (*Paesia scaberula*) lives in the open and is often seen on clay banks along roadsides. It forms a dense, lacy mass and has been used to revegetate road cuttings.

ORCHIDS OF OUR COAST

Seeing that plant *move* really did it for me! I had read about the way the greenhood orchid (*Pterostylis banksii*) entraps insects into pollinating its flower, so I set out one day to verify this for myself. In springtime on the edges of tracks in our forest we find dense patches of greenhood orchids, which sprout from underground tubers. I located a greenhood in a patch of sunlight. Maori call it tutukiwi – 'standing kiwi' – because the elongated petal extending the curving greenhood recalls a kiwi's long beak.

I nestled prone on the damp ground and approached with my lens. I inserted a twig, pretending to be a visiting gnat and touched it very gently on the labellum or lip, a narrow platform at the very centre. Flick! Hinged at the base it jolted back against a delicate white column bordered by curving wings and lacy white fringes. An insect would have been thrust against the plant's sex parts, stigma and anther, delivering pollen and removing that of the orchid. After an interval the love trap is reset.

In an elbow of a pohutukawa on an out-thrust of rock up Ngunguru River snuggles a large clump of immaculate white orchids, pink and golden at the centre. Ladies' slippers (*Winika cunninghamii*) always live in trees. Flowering in January, they are epiphytes or perching orchids, along with the grass orchids.

From the boughs of a big old tree up at the Scow Landing hang tresses up to a metre long of bamboo orchids (*Earina mucronata*), with leaves like fine grass. From October to February they are festooned with creamy yellow and orange blooms in swaying chains. Sometimes they grow on rocks.

At Brattys Bush in August the leaf litter on the forest floor erupts with tiny rubies, rewards for kneeling orchid fanatics: helmet orchids (*Corybas rotundifolia*). On our car park in early spring, high on a windswept ridge, we find a rare relative: *C. cheesemanii*, which resembles a beetle's wing.

On a sunny roadside clay bank in October sprays of dazzling white sun orchids (*Thelymitra*) wave in the breeze. Along their stems rows of up to 20 green buds swell with promise … but it is a stroke of luck to be present when they open. Sometimes they are pink. Some species are cornflower blue. They have a gentle fragrance.

WILD ORCHIDS ON THE LAWN

So far Jan and I have located about a dozen species on this coast – four on our car park! They are often tiny, evanescent and very challenging to photograph. One can always hope to improve. Last year we had masses of orchids of three species in patches on our biggest lawn: the sun orchids *Thelymitra* and the onion orchids *Microtis* ('little ear'). I avoided mowing there until they had seeded. Reward: even more this year – and spreading everywhere.

Each year, too, we get very large greenhood orchids in dense clumps along certain of our partly shaded tracks. And in December, as the other orchids are developing their seed capsules, dark green stems begin to bud. Out will come the weird horned orchids (*Orthoceras novae zeelandiae*). There are three on our car park. Another excuse not to mow!

OPPOSITE, TOP TO BOTTOM: Side view trio *Pterostylis banksii* along our track edges; a greenhood orchid face on; ladies' slipper orchid *Winika cunninghamii;* bamboo orchid *Earina mucronata,* an epiphyte on trees or clinging to rocks.

TOP TO BOTTOM: Helmet orchid *Corybas cheesemanii,* like a beetle's wing; horned orchid *Orthoceras novae-zeelandiae;* a sun orchid *Thelimitra longifolia* opens after midday; common onion orchid *Microtis unifolia.*

NATIVE HERBS

PLANTS THAT EAT MEAT

When Charles Darwin found out that humble little sundew plants could trap insects and digest them he was over the top with delight. In 1860 he wrote: 'I care more about *Drosera* than the origin of all the species in the world.'

I was just as pleased when I began finding these carnivorous plants living in certain locations on our coast. Not in deep forest or well-covered areas; the diet of these tiny plants enables them to forgo rich soil and grow in the toughest, least fertile of places: on tracks, roadside clay banks, eroded sea cliffs, even recently graded farm tracks.

In October the sundews advertise their presence with tiny pink or mauve flowers. Suspended from the tips of a circlet of tentacles are minute translucent globes of sticky gel. If an insect alights in quest of sweet nectar the gel entraps it like flypaper and the tentacles fold slowly to the centre. Enzymes in the gel coat its body. The plant then digests its prey, breaking it down into a nutritious goo that it can absorb. Darwin experimented with bits of egg or meat. He could even activate a sundew with a hair from his head. A raindrop, however, brought no response.

LILIES THAT GROW IN TREES

Characteristic of rainforests are all the epiphytic plants that cling to the trees, carpet and entwine their trunks or perch in their forks: ferns, vines and even small shrubs. Most noticeable on our coast would be the massive perching lilies (*Collospermum hastatum*). In fierce storms their heavy aerial clumps may crash down. 'Widow-makers', old-timers called them with gallows humour.

Most people would assume these lilies are feeding parasitically from the tree, but this is not so. With their three-ribbed, spear-like leaves perching lilies collect and amass falling foliage from the forest canopy and store it in their spongy root base. They actually create a rich mass of soil up in the fork of a tree. Gardeners know windfalls provide a perfect potting mix. From high summer until autumn the striking yellow flower spikes appear. Related are several species of rock lily such as the delicate rengarenga (*Arthropodium cirratum*) of the rocky coast and the most impressive, spring-blooming Poor Knights lily (*Xeronema callistemon*), which coastal people love to grow in pots. But they may have to wait up to 15 years for it to produce its first handsome flower spike.

Scientists group many of our small, non-woody plants under the generic name of native herbs. Because they can't compete for light with the rainforest, so many of these plants are found at the margins. Along tracks and forest edges

ABOVE, TOP TO BOTTOM: Sundew *Drosera auriculata*, a native carnivorous plant; beautiful but deadly, the sticky insect-catching stipule of sundew.

OPPOSITE, TOP TO BOTTOM: Rengarenga rock lily *Arthropodium cirratum*; Poor Knights lily *Xeronema callistemon*; forest edge native iris *Libertia ixioides*; sand dunes native ice plant *Disphasma australe*.

I have found our native irises; where mangroves meet the bush, beds of golden bachelor buttons; on the crests of sand dunes, succulent pink and yellow native ice plant; on coastal rock outcrops, our pristine white tiny native lobelia; along streambeds in the tall forest, the stunning bronze, mauve and green, heavily veined foliage of parataniwha or native begonia, whose serrated, shade-adapted leaves catch the muted sunlight like shot silk.

All these treasures of nature, and so many more, are the reward for exploring the coast in every season and across a wide range of locations.

GIANT GRASSES, SEDGES AND RUSHES

All the world over no group of plants could be more useful. Because they provide abundant fibre and seed, grasses, rushes and sedges have been used by humans for millennia for food, for clothing, for shelter, for rope and nets, and for sails. Remnants of grasses appear in fossilised dinosaur dung.

RAUPO

Known as bulrush in England and cattail in America, the raupo (*Typha orientalis*) growing in the swamps along our coast is a fine example of a tall wetland grass. For the Maori few plants were more versatile: pollen from the chestnut-brown flower spikes was baked into rich yellow bread; fluffy seeds were used for bedding; the strap-like leaves were dried and woven into canoe sails, toy kites, walls and thatched roofs; they were fashioned into hats and the poi used in rhythmical music, and into baskets for transporting moa eggs. The stalks with their structured air cells made buoyant rafts. The starchy roots or rhizomes were baked and eaten.

TOETOE

In marshes but also along water weeps on hill slopes, such as at our favourite Cabbage Tree Bay, are clumps of the tall, graceful toetoe grass (*Cortaderia richardii*). It is often mistaken for the South American pampas, but toetoe has a drooping seed head while pampas is stiff and erect. Toetoe flowers in late spring; pampas in autumn. Toetoe has creamy white plumes; those of pampas have a pinkish tinge. Toetoe is endangered. Pampas is an accursed weed.

OIOI

On the rocky coast in brackish tide pools above high water, and beyond the mangrove glades in broad salt marshes at the uppermost catchment of our tidal rivers, dense swathes of jointed rushes or oioi (*Apodasmia similis*) ripple in the breeze. Their colours range from olive green to orange brown in a flowing tone poem. Among them nest and forage the most secretive of our native birds: the fernbird and the bittern; the banded rail and the spotless crake. The last I have yet to lay eyes upon. The others, but rarely and with immense delight.

SEDGE

Along tracks and in forest clearings the sedge (*Gahnia lacera*) grows with sharp-edged leaves and shiny rice-like seed heads; it is also called the bush rice plant or cutty grass. It is host to the caterpillar of our endangered forest ringlet butterfly. In vain, far and wide I have searched for a sedge leaf with tell-tale signs of grub munching. Exotic predatory wasps are the problem.

In today's world raupo, toetoe, rushes and other swamp plants should be regarded as the kidneys of our land use systems: they have a huge capacity to filter out toxic heavy metals from our roads and excess nutrients from our farms. But for all this immense value, 80 per cent of our swamplands have been drained. Without its kidneys the land bleeds toxins to the sea. Since the 1920s toxic marine algal blooms have become a problem, rendering shellfish inedible. They have been linked to increased nutrient run-off.

TOP TO BOTTOM: Raupo seed head *Typha orientalis;* toetoe flowerhead *Cortaderia richardii*; jointed wire rushes oioi *Apodasmia similis;* sedge 'cutty grass' *Gahnia lacera* seed head.

OPPOSITE, TOP TO BOTTOM: The stinkhorn *Aseroe rubra* mimics dead meat to attract flies; basket fungus *Ileodictyon cibarium*; devil's claw *Anthurus archerii*; toadstool *Cyptotrama asprata*.

FUNGUS 'THINGS'

The first time I ever saw a stinkhorn I thought I was hallucinating. When I began taking pictures on the forest floor along this coast, after many years of underwater exploring, I soon noticed there are fungi that resemble corals and sponges under the sea. This became an exciting theme for me as a diver; a link. But then, among the leaf litter, I saw *a sea anemone*! Pink tentacles radiating from a flat disc, standing erect on a white column. As I focused in on it I sniffed a peculiar odour of rotting flesh. A blowfly began to circle, homing in on the smell. It alighted on the edge of the saucer and began probing at the viscid brown gunk within.

The stinkhorn or flower fungus (*Aseroe rubra*) is a brilliant mimic. It secretes a substance so like putrid meat that flies are totally deceived and attracted to dabble in it. The fungus accomplishes its purpose in life: the fly carries away its reproductive spores and it is spread through the forest.

Later I found a variation on this theme: a white sphere that contained a folded-up package, much like a piece of research apparatus landed on the surface of Mars. It bursts open. From within a spaghetti-like mass of white tubes begins to unfold and expand until an oval geodesic structure lies there like the frame of an inflatable tent. On the inside of the struts I see a brown gooey coating: more pseudo rotten meat or dung. The basket fungus (*Ileodictyon cibarium*) has another way to seduce flies into spreading its spores.

Then my neighbour Henry called me over to his woodpile. Where sawdust and woodchips had been accumulating for years, Henry had an array of starfishes: grotesque-looking things on stalks with five red arms around a hollow cone. The arms were smeared with malodorous phony meat: another variation on the fly-deceiving mimicry. These fungi are aptly named devil's claws (*Anthurus archerii*).

With the first rains of autumn a wonderland of fungi appear, all fruiting bodies arising from underground structures that feed on plant roots or permeate the tissues of dead trees. To achieve their purpose I find our local fungi take so many amazing, often beautiful forms. Some are not so lovely. Toadstools shed their spores from gills beneath a disc or cone; bird's nest fungi wait for raindrops to plop into their minute cup-like structures and splash out their sticky egg-like offspring. Some so resemble brightly hued undersea sponges and corals I have slipped a picture of one in with underwater pictures and my diving mates did not notice the difference!

THE DAY OF THE TRIFFID

One January day, just before rain, a very odd fungus developed at the shaded base of an old but quite sound tree stump near our house. Initially it was a small, bright sulphur-yellow patch with isolated blobs out from the border. I took photos of it. Jan and I have a good collection of local fungi and lichen pictures and had never seen the like. It preceded a prolonged dry spell.

Later that day the bread-like yellow mass travelled sideways along the stump base, overwhelming a moss patch and ensnaring a few grass blades in its tissue. That afternoon, as I was removing grass to better expose it for more photographs, I accidentally touched it lightly. It was sticky and smelt repulsive. I washed my hands well. Ants on the trunk were avoiding it.

Next morning, to our consternation, it had vanished completely from around the stump base. Not a trace remained, even amidst the moss. But a weird, swollen ribbon of it had migrated about a foot up the stump in a horizontal bulge. It was about 40 cm across, like half a baguette of French bread. It seemed like a wave advancing up the stump. It had to be in motion, but this was not discernible. I took more photos. When the first light rain sprinkled it, a brown crust like dried blood developed and it gradually shrivelled to a dark ribbon that persisted for some time. With research, I found it is a species of slime mould. I cancelled plans to evacuate the area.

The slime mould is not truly a fungus, but a mass of naked protoplasm that creeps like a giant amoeba, flowing over surfaces, digesting dead leaves and wood as it goes. Some species are said to glow in the dark. Another is aptly called 'dog-vomit slime mould'. If slime moulds increased dramatically in size I think I would shift out …

Yellow slime mould moving up the tree stump.

MOSSES, LIVERWORTS AND LICHENS

These are the spangles and gems; brocades and magic carpets that adorn our forests and our coastal rocks. Mosses, liverworts and lichens are so old, many of them belong to the ancient continent of Gondwana and may date back long before dinosaurs, some 320 million years or more, having evolved from aquatic plants that came ashore.

MOSSES

Mosses have spongy masses of fine rootlets, but with these early plants roots serve mainly to anchor rather than nourish. They gain most of their moisture through their leaves. Unlike later plants, the leaves have no pores: they just absorb through the cell wall. Dainty in structure, the narrow leaves are only one cell in thickness, bearing a single vein from base to tip. Such plants offer little of substance for the fossil record, so their past is rather vague. Club mosses have more substance. In the Carboniferous period they grew as big as trees and became part of the coal deposits. The tree club moss (*Lycopodium deuterodensum*) we find along tracks on our land can reach 600 mm high and resembles a tiny Norfolk Island pine.

For the most part mosses live in damp, humid locations and often form

Mosses degged with dew resemble coral polyps.

dense clumps to retain moisture. I have always enjoyed their resemblances to undersea bryozoans (moss animals) and coral colonies: both land plants and sea creatures grow with pleasing symmetries, so that the forest floor often seems like a green coral reef to me. Some, like the pincushion moss (*Leptostomum inclinans*), are epiphytes and live on the trunks of trees. Mosses reproduce with spores shed from a terminal capsule, as in the picture [opposite top left], or by simply splitting off a tiny clone.

LIVERWORTS

Closely related to the mosses, but with subtle physical differences, the liverworts often have irregular lobes. Others resemble mosses so closely the difference is academic. With our wet temperate climate New Zealand has the greatest density of liverworts in the world. I find so many weird shapes and designs of mat-forming liverworts embroidering tree trunks, rocks and damp, very shady places, it seems like I am exploring science-fiction life forms from another planet.

LICHENS

Weirdest of all are the lichens. Under the sea scientists discovered that the coral polyps that form massive reefs live with tiny plants in their tissues. The anemone-like animal shelters microscopic algae and they in turn provide nutrients from photosynthesis for the coral. A close parallel exists with the lichen. It is really a special kind of fungus that encloses and protects green algal cells in its tissues that can fix nitrogen from the atmosphere and create sugars for its nutrition. Lichens are superb survivors of climate extremes and severe desiccation. They thrive in the most demanding of locations – such as the wave-splashed, sun-baked rocks of our coast, tree trunks or rocks in the Antarctic. They cover the planet from polar wastes to sandy deserts, and have even survived a couple of weeks in the vacuum of space, orbiting Earth.

Tree trunks and boughs may be richly clad with lichen of so many kinds: from a powdery yellow, almost fluorescent coating to the luxuriant flounces of the foliose lichen *Pseudocyphellaria billardierei* that shrivel during dry weather and expand with the first rains.

On tree stumps in April pixie caps of the lichen *Cladonia floerkeana* develop pretty bright-red knobs. Then I see an array of pale-green cups: spores develop in these receptacles especially shaped so that raindrops splash into them and spread them.

CLOCKWISE FROM TOP LEFT: Emerald pincushion moss *Leptostomum inclinans* with terminal spore capsules; lobed liverwort *Marchantia* sp.; foliose lichen *Pseudocyphellaria billardierei* on a trunk; pixie cap lichen *Cladonia floerkeana* – the red knobs develop into splash cups full of spores.

Kauri forest at dusk.

CHAPTER 7:

A WILDLIFE EXPLOSION

Life has become very exciting at Riverlands. It seems our Landcare Group is on a journey back to Gondwana. For two years we and our neighbours have done intensive pest control on rats, stoats and possums. Hey presto! All sorts of unusual native animals are reappearing: creatures whose status is threatened, from birds to lizards to kauri snails and insects, are surviving and breeding. Native plants are proliferating because the rats are not eating their seeds and there are more birds to spread them. Pest control has its ugly preoccupation with killing exotic animals. But the upside is witnessing the real New Zealand emerge: the ancient flora and fauna that have been clinging on to existence on our lovely Tutukaka Coast, just waiting for a chance to flourish anew.

LAND CRITTERS

OUR NATIVE GECKOS

Whenever Jan goes into a small sleepout in our forest to clean it for visitors she finds them: the shed skins of geckos, complete in every detail, including tiny feet and toes like gloves. Brown, white-tipped droppings litter ledges and furniture; even sticking to the walls. When she examines them she notices body parts of the inoffensive native cockroach. So this is part of their diet. Because I built this sleepout to be rodent-proof it has become a gecko haven. A visitor sleeping there had a gecko drop onto her bed and scurry away. These have been Pacific geckos – quite rare in most places because of predation by rats and mustelids.

Gradually, as our pest control improved, we began finding the night-hunting geckos in all the outbuildings on our land. A Pacific gecko clung to the insect screen on our ranch slider, peering into our lounge. During a drought another unfortunately drowned in a bucket of water. In the tropics geckos live in almost every room. A jocular friend of ours once complained to his hotelier in Port Vila because his *chambre* had no gecko …

One day I lost my wristwatch while working on our long bush-lined driveway. Next day I walked down feeling how hopeless it would be to find it. Suddenly my eyes were drawn to an emerald green creature perched on some sticks at the base of a blackwood tree. A green gecko! On the ground! And right beside it lay my watch.

Pacific gecko
Dactylocnemis pacificus.

Green geckos are even rarer than Pacific geckos in our district. I have seen these day feeders five times in 40 years and each time have felt blessed. Twice they were in their typical habitat: among the crests of manuka where their gold-

Auckland green gecko *Naultinus elegans.*

splotched bodies appear like sun-dappled foliage. And I got a photo of that one.

With one exception all New Zealand's geckos (more than 40 species) and skinks (over 50 species) produce live young – unlike most of those elsewhere in the world, which lay eggs. Our geckos usually produce just a pair of young, which makes them particularly susceptible to exotic predators. The young appear around April and are surprisingly big; about half the size of the mother. They mature in about three years. Like crayfish, with growth they shed their skins periodically. Geckos are also long-lived, reaching 10–15 years in the wild, and longer in captivity.

As anyone who has lived in a tropical hotel room would know, geckos are very vocal, with chirps, barks and clicks. While they have no eyelids they are able to clean the eye screen with their long tongue.

Both of our geckos are omnivores and eat a wide range of insects, caterpillars and small spiders as well as small berries such as those of coprosmas and mahoe and the nectar from flowers such as flax and rata. The Pacific gecko is a nocturnal feeder and loves moths. It is able to match its colour pattern to the background. By day it basks invisibly among leaves or on bark. The green gecko is diurnal. It feeds by creeping up on insect prey and making a rush to seize it with small sharp teeth in a wide set of jaws.

Beside their native predator, the kingfisher, geckos suffer severe predation by exotic pests such as rats, stoats, cats and possums. The last also compete with them for certain foods. In our neighbourhood those pests are on the wanted list.

TOP TO BOTTOM: Ornate skink *Oligosoma ornatum*; kauri snail *Paryphanta busbyi* withdrawn.

OPPOSITE, TOP TO BOTTOM: Kauri snails mating after rain in May; a kauri snail travelling; a kauri snail devouring a native worm.

THE TATTOOED LIZARD

Life has its odd resonances. The very next day after reading about the protection of the rare, 'chronically threatened' and very secretive ornate skink (*Oligosoma ornatum*) on Limestone Island Reserve in the Bream Head Conservation Trust newsletter, we got photos of one next door on Neal and Reggie's Riverlands lawn.

We were seeking drone helicopter photos from 120 m above Ngunguru River on a morning of perfect calm. Operator Colin Gans spotted it scurrying through the close-mown grass. And it stopped dead still. Two cameras whirred. A new lizard to add to our two gecko species here. We have never seen one in 36 years of occupation. We read that newsletter. Next day: woweeeee! The ornate skink seems, from head on, to have a Maori tattoo on its face – a moko – and a pale-yellow teardrop beneath each eye.

Reggie maintains an immaculate woodpile along his boundary with our forest. He has told us of seeing something scurrying for cover when he fetches wood. The stack must be a deluxe lizard high-rise right beside our covenanted forest and adjacent to a hillside of unmown grass.

Like our geckos, New Zealand skinks are especially vulnerable to predation. They are live bearers (but for a single South Island species that lays eggs) and may produce up to eight young; there is no parental care and the offspring must be self-reliant.

KAURI SNAILS ON OUR LAND

On a day that brought the first light rains after five months of drought, Jan and I cleaned out the side channel on our driveway ready for heavy rainfall events in the coming winter months. Removing leaf litter accumulations we found 16 live kauri snails. Only two were juveniles; the largest was almost seven-cm across; large, glossy, olive-green creatures, gleaming like forest jewels and as elegant as any spiralling seashell. Oddly, under camera flashlight they appear a rich brown.

Kauri snails are a threatened species, not normally known in this district. They suffer heavily from predation by rats and pigs, and they are trampled by stock. We have been finding them on our land for several years now, ever since I released one into our bush that a pig hunter brought to show me some 15 years ago. He had found it up north near Whangaroa after heavy rain. (They were being crushed on the route, but removing the snail was an unintentional illegal act!) In wet weather kauri snails tend to become more visible.

The translocatee must have had eggs and liked the habitat on our land. Jan and I have since found them as close as one metre from our house and they have spread far and wide through our 18-hectare forest-covered property. Once, when I was clearing a step in our bush, I rolled back some moss and leaf litter to reveal a pair of large ones a few inches apart. They were a considerable distance from the point of release. Now kindly neighbours even stop their car at night to avoid them on their drive. Year by year we find kauri snail numbers increasing. But rat control has been a major effort for us. Kauri snails have found this place a paradise and we feel so grateful for the reward they provide for our hard work and expense setting baits and clearing traps of rodents day after day. Since introduction the population of these little carnivores has certainly boomed here; a valuable indicator species.

Exploring several tracks in the Tangihua Mountains, a major DOC reserve west of Whangarei, Jan and I were concerned at the frequency of wild pig rootings and the abundance of shattered kauri snail shells. It was like walking along a beach littered with shells, crunching their remains underfoot. Kauri snails should never be pig food. Wild pigs are creating havoc along these tracks and must be hunted down if the forest community is to be sustainable.

From research into their lives I learnt why we see so many kauri snails around after heavy rain. Climatic events can trigger mating. In the wet, these hermaphroditic molluscs lie connected for up to 10 hours or more. Among leaf litter they lay nests of dull white, shelled eggs, each about a centimetre long. Juvenile hatchlings initially seek shelter in the branches of trees and shrubs, up to 6 m above ground. Paradoxically kauri snails do not favour areas near kauri trees, as such spots are often too dry.

Maori did not eat kauri snails or pupu rangi. The snails predate earthworms, insects and their larvae, slugs and other snails. They feed only at night. They

pose absolutely no danger to garden plants and other vegetation. Plants would benefit from their predations. The shells of our species, *Paryphanta busbyi*, can reach 79 mm in diameter and they can live 20 years or more. It takes seven years for one to reach 54 mm. (There is another, slightly smaller species in the far north, *P. watti*, that reaches 69 mm and about 18 other species elsewhere in New Zealand, 17 of them endemic to the South Island.) But all kauri snails are now endangered and those in Northland live in small, scattered colonies where they have been saved from extinction by the efforts of DOC. So I now realise that, inadvertently, we have done the handsome molluscs a great benefit by controlling rats and other pests, and eliminating stock. This has enabled them to breed prolifically. They have now spread to our neighbours' properties, much to their delight. A cyclist who passes our place regularly tells us they have become roadkill.

Like the weta and tuatara, kauri snails are living fossils from Gondwana, belonging to the oldest land snail family, which originated 200–300 million years ago in the late Palaeozoic or early Mesozoic.

GIANT BULLY

I thought I knew my neighbours; that is, the local wildlife was pretty familiar to us after years of study. So Jan and I were staggered when neighbour Hilton Ward, who runs the River Homestay by the Crawford Reserve, rang us one spring morning. 'I've just been cleaning a drain from the mangrove swamp into Waiotoi Stream. By the floodgate I found about a hundred fish that look like small groupers.'

On investigating it turned out Hilton was dead right. By the mouth of a tidal creek that drains from the forest into the Waiotoi, a major tributary of the Ngunguru River, he caught a couple for us to photograph and film in our aquarium tank. They were quite big: 15 cm long, with protruding jaws and large mouths fringed with many small, sharp teeth. They were like miniatures of the big tropical grouper I have seen lurking in grottos on coral reefs. The similarity is probably no accident: by convergent evolution these lurking predators, the huge grouper and the tiny bully, have developed a similar form for the same purpose. The shoal of small fishes hovering in the creek nearby, first of the season's whitebait (or galaxiids: juvenile native trout), may have been the reason for this aggregation of pygmy sea wolves. Or they may have been seeking breeding grounds.

From its protruding lower jaw, its fin patterns and a count of fin rays, this is the giant bully (*Gobiomorphus gobioides*), a freshwater fish of the worldwide family Eleotridae. New Zealand has seven species of eleotrid bully, but at up to 25 cm this is the largest. Giant bullies live in coastal areas, just above the tide, all around the country. They feed mainly at night on small animals such as insects and crustaceans.

Their breeding habits are very similar to those of demoiselles that we have studied in the sea. Males set up territories and defend them vigorously at breeding time, when their skin colour tends to darken. They court females that are ready to spawn. As she deposits her eggs in a hollow under a rock, he fertilises them. Now, just like the male demoiselle on the coast or at the Poor Knights Islands, he must defend them from every predator until they hatch. Larval bullies go out to sea where they spend their initial months, so bullies can only live in places accessible to the sea.

Once they had been photographed for identification, and recorded on video, our giant bullies were returned to their point of origin, none the worse for wear. Later, I heard on the radio that about 70 per cent of our native freshwater fishes are endangered. Farm run-off is the main culprit.

KOURA

By day, beneath the shady banks of our small stream, hides a relative of the ocean-dwelling lobster of undersea caves. With its twin set of powerful pincers it resembles a miniature Atlantic lobster. At night the koura crawls out to feed. Since

Koura *Paranephrops planifrons*.
ALISON PERKINS

it scavenges old leaves and organic detritus, water conditions are enhanced for other stream dwellers. Because of this, the koura has a special importance: like the native pigeon it is regarded by ecologists as a keystone species, and the presence of koura in our forest stream is an indication that all is well.

Slow growing, the North Island koura takes three years to reach maturity at a length of 2 cm; fully grown, it measures only 7 cm. Like the crayfish of our diving world, at each stage of growth this crustacean has to shed its protective shell. Taking water into its tissues, it expands a little and then a new shell hardens into a suit of armour. Over its lifetime the koura moults like this many times, nine in its first year alone. While in the soft shell state it is especially vulnerable to attack. When threatened the koura can reverse swiftly with a powerful flick of the tail.

The female produces eggs in autumn and again in spring. May–June is the more important breeding period. The eggs are carried under her tail for about four months until the juveniles hatch, each a tiny replica of the adult. Until the young are about 4 mm long and can fend for themselves they cling to their mother for safety.

Unfortunately the koura is now a threatened species. Its habitat – swamps, ponds and other freshwater places – is gradually declining. Introduced trout compete with it for aquatic insect life and eat its young, and overharvesting by humans is an added factor.

PURIRI MOTH

Whenever I enter my favourite tree, that hollow puriri on an exposed headland not far from our home, I gain new respect for New Zealand's biggest moth. This tree is healthy and strong enough to withstand cyclonic winds. It has been gutted by the larvae of a beautiful bright-green moth that feeds on its sapwood. The moth's colour, with subtle markings like Maori carving, matches the leaves of the tree in which its life begins.

With a wingspan of 15 cm the puriri or ghost moth (*Aenetus virescens*), known to Maori as pepetuna, is our biggest native moth. It lays it eggs on a host tree where the larva burrows and feeds for up to five years, by which time it is about 10 cm long. Then it forms a golden, white-banded chrysalis attached to the tree bark and transforms into a moth. The adults have only two days to reproduce and lay eggs. They have no mouthparts for feeding. At dawn and dusk on fine days in springtime, from September to November, they are active in quest of mates.

A neighbour found the moths were attracted to lights on her patio, blundering about clumsily, often damaging themselves. When she picked one up to examine, it was suddenly plucked from her fingers. A morepork had snapped it up on the wing. This was repeated and soon became a party act. The moth is also food for other native birds such as kaka and robins and also, nowadays, for rats and cats. Maori ate the larvae or used them for eel bait. Larvae also feed on manuka, lancewood, wineberry, beech and titoki; on introduced trees such as apples, oaks, poplars and willows; and hardwoods such as eucalypt and Tasmanian blackwood. Small-diameter trees or branches may be ring-barked and die.

NATIVE CICADA

There are 42 species of cicada in New Zealand. The commonest and largest – up to 14 cm long – is the chorus cicada (*Amphipsalta zelandica*). For months these bejewelled sap-suckers cling to every bush, shrub and tree on our land, engaged in some sort of sonic territorial battle or sex display. Only the males sing and many species have distinctive songs. I shudder involuntarily when one lands on me or crashes into my face, but they do us no harm. Gardeners have reason to curse them.

After mating the female stabs plant stems in a herringbone pattern with her saw-like ovipositor and inserts a row of eggs. She does this repeatedly on several plants. Adults live only one season. But the eggs hatch into twin-legged nymphs, which drop to the ground and disappear. Those front legs can burrow as deep as a metre and the nymphs suck root sap, moving around and growing for three to five years until, after shedding their skin about five times, they reach adult size. Then the pale, drab nymph climbs up to the daylight, ascends a plant and makes its final moult: a bright-coloured membranous winged adult emerges. In springtime on every tree and shrub the old sheds, pale brown and translucent, cling with a split in the back where the adult emerged for its singing career. They live only two to four weeks.

Cicada song is made by an elaborate acoustic organ or tymbal that buckles

OPPOSITE, TOP TO BOTTOM: On windy hilltop, hollow host tree of puriri moth; puriri moth chrysalis; puriri moth *Aenetus virescens*.

BELOW, TOP TO BOTTOM: Chorus cicada *Amphipsalta zelandica*; an empty cicada nymph shed; a cicada infected by fungal mould.

and pops in rapid pulses, rather like a tin lid buckling in and out. Special air sacs in the male's abdomen act as a resonating amplifier. By twitching it he can modulate the sound, which seems loudest during the hottest part of the day. Females trace the sound to locate the singing male. Males are able to determine the proximity of rivals. An attractive feature of the cicada is the three miniature eyes or ocelli, spaced between its normal eyes like gems.

Cicadas are food for many birds, including the kiwi, which eats the nymphs. Sometimes they are infected by a fungus while underground.

PRAYING MANTIS

Living on the forest edge, on the tops of shrubs and bushes, it closely resembles a long, green leaf. But the praying mantis is a nimble, highly skilled and voracious predator. The ancient Greeks named it mantis, 'the prophet', because the upraised forelimbs suggest a human figure praying. Ironically *preying* is what is on its mind. It can catch insects as large as itself and snatch flies by the score with lightning movements of those forelegs, which are equipped with impaling spines.

The New Zealand mantis (*Orthodera novaezealandiae*) is nowadays in danger of being displaced by another species – paler green and slightly smaller, with a plumper abdomen – that has arrived here from South Africa: *Miomatis caffra.* The invader was first noticed in Auckland in 1974. Our mantis females reach 45 cm in length, while males are only 35 cm. In rare situations, when the pregnant female is under dietary stress, she may devour her mate as a food supplement.

Following autumn mating (February–April) mantis egg clusters adhere to the weatherboards of our house: pale, crusty symmetrical shells shaped rather like the old-time smoothing iron. These are extruded gradually, over several hours, from the female abdomen, leaving impressions on the upper surface. From them, with the warming of spring, up to 70 miniature mantids will emerge, wobbling around on the walls and entering our windows. After moulting six times over the ensuing months of summer they grow to maturity, reproduce and usually die off before the next winter.

Mantids can fly. Beneath protective wing covers they have gauzy wings. Their diet includes many unwelcome garden pests: blowflies, wasps, moths, cockroaches, and others. Wise gardeners encourage mantids among their plants, but pesticides kill them.

OPPOSITE, TOP LEFT: Praying mantis *Orthodera novaezealandiae.*

OPPOSITE, TOP RIGHT: Stick insects *Clitarchus hookeri* mate, male above.

CENTRE AND BOTTOM: Knobbly stick insect *Acanthoxyla* dorsal view; side view.

STICK INSECTS

I had a soft heart towards stick insects, so very abundant around our home in the forest – until I discovered they were the culprits eating every new bud on our

pohutukawa. The poor tree could not grow a jot. At every attempt, the terminal bud was nipped off neatly. I thought about it ... There are huge numbers of stick insects in the area. Could it be them, feeding in the night? Their camouflage is so excellent that birds would seldom be able to distinguish them from a twig. I steeled my heart and began removing the bud eaters. They were most certainly the culprits. I even spied on them with a torch. That tree is now five metres high – and we still have plenty of stick insects.

There are some 22 species of stick insect in New Zealand and most of them share a common trait: the females do not always need the smaller-sized, skinny males for reproduction. Some, like our smooth, green or brown *Clitarchus hookeri*, can reproduce successfully for five generations without any male input. A researcher found eggs from a mated female gave rise to both male and female offspring in equal numbers, but eggs that developed without a male, that is, by parthenogenesis, produced only females. And without sex, eventually the young become smaller and less in number. This species can grow to 10 cm. But with the prickly stick insects of the *Acanthoxyla* genus, seven out of the eight species always go it alone. There are absolutely no males. These insects, always mothers, usually have black-tipped spines on their bodies and may reach 11 cm in length.

Stick insects are flightless, harmless, nocturnal vegetarians and the various species tend to specialise in certain trees and plants. Their protective green or pale-brown disguises enable them to blend into dense green or brown twigs and bark to survive undetected during the dangerous hours of daylight. If disturbed, stick insects may fall to the ground and play dead, their legs in odd positions. They also manifest a strange swaying dance, which may go on for hours; its function is as yet unknown – perhaps they are trying to mimic the effect of wind. Besides birds, they are preyed on by wasps and possums.

With the limitations of their lifestyle, flightless, silent and confined to the forest canopy, finding a mate would be difficult. Like many insects the female emits a pheromone to attract the male. In spring and summer we see him riding on her back, his abdomen twisted down her side to reach her genitalia. He may remain like this for days while they continue to walk around among the foliage and feed. From time to time she drops batches of fertilised eggs that fall to the ground.

After about three months or longer, lying like small flakes among the leaf litter, the eggs hatch and tiny, long-legged replicas called nymphs totter about. At this stage they are all green, but after the second growth moult some may become brown. Size, not colour, indicates stick insect gender. They moult four times to reach maturity and can live up to two years. One species, *Argosarchus horrida*, can reach 22 cm in length – yet another New Zealand giant!

WETA

I almost fell out of that tree. During a filming expedition at the Poor Knights Islands 40 years ago I met my first giant weta (*Deinacrida fallai*) at very short range. I had climbed up to check out a bird nest. As my eyes reached the level of the nest I found myself staring into a pair of coal black, bulging, compound eyes bracketing the face of a monster. On the film sound track Dr Walt Starck described it aptly: 'A face only a mother could love.'

Little did I know I would later live in a house surrounded by weta. There would be weta crowding our basement walls; weta yearning to enter our windows at night. In the trunks of manuka around our home weta would live in neatly drilled holes formed by the larvae of huhu beetles or puriri moths and serenade us with their night noises. But even though their name in Maori means 'god of ugliness' and, as one of the world's largest insects, they may well inspire terror, I have never known anybody in our weta-filled part of New Zealand to be hurt by one. That is, unless they fell from a tree.

Weta are flightless nocturnal insects, related to the grasshoppers and are living fossils older than the dinosaurs, little changed in the past 190 million years. Besides the giant weta of the Poor Knights Islands (up to 200 mm long) and the even larger species Jan and I met one night on nearby Little Barrier Island (*D. heteracantha*), which resemble rats or small crayfish, we have two species in our neighbourhood: the tree weta in our forest and the cave weta that inhabit our neighbours' lakeside pumphouse. In all, New Zealand is home to some 70 species.

Female tree weta *Hemideina thoracica.*

They have a complex social life. A dominant male lives with a harem of about 10 females and about as many young. The male tree weta develops an enlarged head and jaws almost twice that of the female; at her rear the female has a long curved ovipositor for laying eggs. (It looks like a massive stinger, but any harm to humans from weta comes from their jaws, used in defence.) When about 18 months old the female lays a dozen eggs which take three to five months to hatch. Meanwhile the parent weta usually die off.

By night, weta emerge from daytime refuges to feed on plant material such as lichens, leaves, flowers and fruit, and they will scavenge dead insects and other forest floor organic waste. Like crickets, they communicate by rubbing their especially roughened thighs on the abdomen to make a stridulating sound. Near the knee joint on their front legs they have hearing organs. If threatened they raise their spiky hind legs as a warning and can strike at a predator or competing male, besides wielding a powerful set of mandibles on the head.

Some of our plants have seeds designed to pass through the weta's gut undigested, hence these insects have a role as seed dispersers. Weta are food for nocturnal hunters such as kiwi, morepork, tuatara and bats. But these days they are threatened by hedgehogs, rats, cats, mustelids and other exotic pests.

Cave weta have relatively small bodies but unbelievably long legs, and their antennae, used for feeling for food in the dark, can be seven times the body length. If they sense disturbance they may jump up to two metres.

Cave weta *Gymnoplectron uncata*.

OPPOSITE, TOP TO BOTTOM: Giant centipede *Cormocephalus rubriceps*; centipede head and pincers; centipede tail and rear antennae.

GIANT CENTIPEDE

'There's something in my hair,' screams Jan, sitting up in bed. I turn on the light to drive away the nightmare. But it is real. A serpenting, multi-legged, dark-brown fantasy has just slithered from her hair onto the pillow. Had I not got rid of it out the window there would have been no further sleep ... Each year with the onset of winter a giant chooses to enter our house. It has always been sent swiftly outside. Only once, when reaching onto a high shelf in the boatshed, my finger was nipped. I reacted so quickly there was no injury; no venom. Life in Northland is far too pleasant to let these giants discourage us. They are everywhere on our land; under every pile of wood or drift of debris. We have to share our space with monsters. They were here first.

New Zealand has some 35 species of centipede in all habitats from coastal to alpine. Most are of average size, but our native giant centipede (*Cormocephalus rubriceps*), weri or hura, may reach 160 mm from its front antennae and pincers to its rear-facing antennae and 10 mm across its segmented, armour-plated back. On offshore islands such as the Poor Knights it can reach 250 mm, but predation by rats limits growth on the mainland. It is a major predator of anything soft bodied that moves on the forest floor: spiders, worms, snails – even small lizards. It seizes prey with claw-tipped pincers and injects it with a shot of lethal venom.

These centipedes are almost blind; they can detect only light and shade. Attached to each body segment is a pair of legs; 21 with this species. They do not copulate; males just leave a package of spermatophores to be picked up by the female. But the female does care for her eggs, lying coiled around them like a Chinese dragon until they hatch, licking them periodically to prevent fungal attack. I have seen one bearing its eggs under its tail.

HUHU BEETLE

On summer evenings the huhu males are rampant in our district. From trees and bushes come enticing smells from the females who put out their sex attractant. Like high-tech warplanes these beetles bristle with sensors: huge compound eyes like radar domes; extremely long, sensitive antennae to detect odour and determine directionality. Their mandibles can inflict a sharp bite. But in their two weeks of existence, adult huhu beetles never feed; the powerful jaws are a redundant feature of their youth. Mating after dusk is their all.

Bright lights really upset their navigation systems; finding an open window they blunder noisily in and bounce off all the walls. It is quite a challenge to get rid of one. Jan was bitten on the finger trying to take these pictures. From November to March is the huhu mating season, with an emphasis during midsummer (December–January) when people live in tents. And they are most active from dusk to midnight, as are the campers they so often terrify.

The huhu or tunga rere (*Prionoplus reticularis*) is a longhorn beetle, and at up to 50 mm in length it is New Zealand's biggest beetle. Larger than the male and lacking his patterned wing covers, the female lays her 3-mm cigar-shaped eggs in batches on the bark of fallen trees. Larvae hatch in up to a month and then, through a 1-mm entry hole, they burrow and chew their way into the wood, creating long galleries. For up to three years the fat white grubs tunnel, excreting digested wood fibre and ejecting it from the tree in pellets of powdery white frass. They grow up to 75 mm in length. At this stage they are a delicacy much prized by early Maori. I have tried one fried. It has a mild peanut flavour with a hint of turpentine.

Once fully grown the grub forms an oval chamber within which it transforms into a pupa. Two and a half weeks later an adult huhu beetle is ready for flight. Those strong jaws are now very important as, through a 20-mm exit hole, it chews its way out. The huhu beetle has an important ecological role as it helps break down and recycle waste wood. But out in the forest world, during the mating season, it is a favourite food for the night-flying morepork. Hedgehogs, rats and mice also eat huhu. A night-prowling cat is said to have died after eating one.

GIRAFFE WEEVIL

The giraffe weevil (*Lasiorhynchus barbicornis*) is the longest, weirdest beetle in New Zealand – and it is not a weevil at all! The male can reach 85 mm in length. Fortunately it was not the male we found on our ranch-slider insect screen one day in late February, but the less daunting female, who only gets to 45 mm. Half of the male's length is his huge rostrum or beak, which has antennae and wood-chewing mandibles at the tip. The female's rostrum is only about one-third of her body length. Males use their extra endowment for fierce battles

with their own gender. For females it is a burrowing instrument for egg laying: they insert it at an angle into dead wood on trees and chew a hole with their terminal mandibles. One has been seen screwing her head from side to side and alternately withdrawing to eject debris until she had made a square hole 4 mm deep. She then reversed, laid her eggs within and plugged it up with bark. As she was making the hole her long-snouted mate copulated with her. Egg laying is from October to March. After two years chewing wood and growing, the grubs emerge as adults and live a further two weeks to mate.

This is when beak battles take place: if a rival male arrives he tries to knock the copulator off with a rostrum blow on the back. If this fails he tries to seize one of the copulator's legs in his mandibles and vice versa. Should either succeed the rival submits and is lifted up and dropped to the ground, where he hides among the leaves.

Amazingly these beetles can fly, albeit rather clumsily, blundering about in daylight hours. Preferred trees for egg laying are karaka, lacebark and pigeonwood, kauri, rimu, pukatea, tawa and rewarewa. Occasionally adults have been seen feeding on tree sap. Amazing that they live for only a fortnight!

OPPOSITE, TOP TO BOTTOM: Huhu beetle head close up; The male huhu *Prionoplus reticularis* has patterned wing covers and is smaller than the female; huhu side view; huhu grub.

ABOVE: Male giraffe weevil *Lasiorhynchus barbicornis*; a female giraffe weevil perching on a lily bloom.

LEAF-VEINED SLUG

Just as kauri snails do no garden damage, our leaf-veined slugs (*Athoracophorus bitentaculatus*) are not vegetation eaters. Extending their sharp feeding radula, they scrape tiny fungi from leaf surfaces in the forest. Their bodies are appropriately camouflaged. New Zealand has some 30 species of these attractively coloured molluscs and some may reach 10 cm.

On the upper surface, near the head, is a special pore: the air-intake to a unique lung structure. This mollusc has adapted to air breathing, although much of its needs are still absorbed through the moist skin. Leaf-veined slugs live in damp protected places such as at the bases of flax and nikau palm leaves or among forest floor litter. Jan found this one on the leaf of a shaded lacebark tree.

TUNNELWEB SPIDER

When she shifted a garden seat to mow the grass under it, Jan just had to run and get her camera. Exposed on the ground was a very big jet-black spider in some sort of nest. It had long spinnerets at the rear of a hairy abdomen. How many visitors have sat on that hilltop seat admiring the lovely distant view of Whangarei Heads and the flights of tui and pigeons, oblivious that they were within centimetres of one of our largest native spiders?

The tunnelweb spider (*Porrhothele sp.*) can exceed 3 cm in length. It lurks at the mouth of a long silken tunnel, snatches passing prey with its large fangs and drags them into its lair. It can inflict a painful bite, but fortunately for our visitors this is rare. Occasionally, during the mating season, males have been found in houses in search of a female. Jan's photo is of a male, according to expert advice. It is conjectured that his front legs are specially modified to grasp the fangs of the female so she cannot kill him when mating. His abdomen is slimmer than hers, too.

In New Zealand we have over a thousand spider species. Major insect predators, they are also food items for many other species. Just beside the lair of this male I watched a silvereye fly in to feed on the tiny spiderlings within a large twig-tip cocoon, on a large hebe bush, inserting her hungry beak into their nursery from several angles.

CLOCKWISE FROM TOP LEFT: Leaf-veined slug *Athoracophorus bitentaculatus*; tunnelweb spider, a species of *Porrhothele*; lair of a tunnelweb spider; bush giant dragonfly *Uropetala carovei*.

GIANT NATIVE DRAGONFLY

For years Jan and I have been lusting to get a really good picture of the giant dragonfly. We see them zooming around the house, hovering for a moment and whirring away. Never a pause. Never a chance for a photo. Until this March morning!

As Jan steps from our shed something prickly grazes her right forearm. *Cicada* flashes through her mind; there are thousands shrilling around our

house. A bird-length insect leads her across the lawn six strides to a broad tamarillo leaf. She pauses to gaze at the huge dragonfly grasping the leaf edge with its elegant black forelegs. Its yellow-banded body rests on the foliage. Do I have time to get my camera, she wonders, backing off gently and dashing indoors. I follow her out with my movie camera. There she is, stooped over the leaf. This insect is unbelievably big. She takes a wide shot, then zooms in for a second one. I steady my video camera on the porch doorframe. Just as I squeeze the trigger it soars up and away.

A tape measure show the leaf is 12 cm across where the insect settled. It must have been at least 10 cm in body length. References tell us they reach 8.5 cm with a wingspan of 14 cm. But a scientist advises me that our dragonfly, with an estimated size of over 9 cm body length, is likely to be *a world record*. Around 300 million years ago there were dragonflies with wingspans of up to 75 cm – until carbon dioxide levels rose, denying them adequate oxygen.

The bush giant dragonfly, kapokapowai or water snatcher (*Uropetala carovei*) is the largest of New Zealand's 11 dragonfly species, a bionic helicopter that hovers in search of insects in forest clearings and along the edge of the bush during summer. It is a voracious predator, observed to catch up to 20 flies in an hour, and to fly as fast as 70 km/h in any direction – even backwards. It lives only to reproduce; couples mate joined tail to tail in flight, and the female lays her eggs in fresh water, where the predatory and aquatic nymph then lives for about four years before moulting into an adult.

DAMSELFLY

A relative of the dragonfly is the smaller damselfly, which hovers around Lake Ngunguru next door. The main difference is the way its wings close over its body when at rest.

For ages Jan and I have been building up archives of wildlife pictures: trying to get all the flora and fauna around the north. At times we wonder, what next? What more can there be? Then nature gives us a good kick. Lately we have been haunting the edge of an artificial lake near home trying to build a fish ladder there to admit native fishes. Surrounded in forest, it is still discharging clear water into the mangroves, despite a severe fossil-fuel-induced drought. And there are so many damselflies – including a stunning little creature only 25 mm long that grasps a stalk in horizontal handstand. It is so exotic in its colour scheme we think we are in the Sumatran jungle: emerald head, orange-red fuselage, electric-blue tail band. With zoom and extension tube, Jan zeroes in on a male *Ischnura aurora*. I video another species, an emerald female hooking her tail around to her head and grooming. A bright red one, too; scientists are still working out its identity at a molecular level.

NEW ZEALAND GIANTS

Our giant dragonfly, along with other local whoppers, is an example of a widespread and common feature among New Zealand insects: a lot of them are very large indeed. Many of them have a common origin in Gondwana, and they radiated geographically before Zealandia drifted away from the ancient supercontinent. And what's more, many of the giants live in the protected forest around our Ngunguru home. Land of the giants! Large weta, kauri snails, puriri moths, stick insects, huhu beetles, centipedes, leaf-veined slugs, giraffe weevils, and 1.4-metre-long north Auckland earthworms abound.

TOP TO BOTTOM: Red damselfly *Xanthocnemis* species; blue damselfly *Austrolestes colonsoni*; Male aurora bluetail damselfly *Ischnura aurora*.

NATIVE BUTTERFLIES

Of the four native butterfly species we see on this coast, one is quite abundant but tiny and inconspicuous; another prefers a host plant more common in certain habitats along the coast; and two are very rare because their host is also rare.

Fluttering close to the ground on summer days is a bluish grey, furry butterfly: the common grass blue (*Zizina otis labradus*). Only about 2 cm across, it is easy to mistake for a moth. Its host is the flowers and leaves of the white clover and related plants. Females are brown.

Occasionally, around our house in summer, we see the common copper (*Lycaena salustius*). But Jan and I have found a lonely little bay in Tutukaka Harbour where these abound. They hover above their host plant, the sprawling, tangled bushes of pohuehue wire vine (*Muehlenbeckia australis*) that grow on shingle or sand near the sea. On the small round leaves we found their tiny grubs while all around us ardent pairs of them were mating.

Most striking of all are the red and yellow admirals (*Vanessa gonerilla* and *V. itea* respectively). They are also very rare, and no wonder: only very seldom do we see their native host plant. In a light well in the forest, just where one might choose to sit down and enjoy a patch of warm sun, the native stinging nettle, rather like a very prickly thistle, occasionally awaits the unwary bottom.

All our Mount Tiger neighbour Elizabeth Amber really wanted was a cup of nettle tea, brewed from European stinging nettles. She gathered a stash of nettles, roots included, from a friend's paddock and established a clump in her organic cottage garden. Then, one day she was delighted to find the rare butterflies fluttering over her nettles. Yellow and red admirals had been adhering their individual, barrel-shaped eggs on the leaves. Nettles are their host plant just as monarch butterflies require the swan plant.

After about 10 days she saw tiny black caterpillars, their backs protected with a covering of branching, golden hairs, crawling about and gnawing at the leaves. Then she noticed some of them were drawing nettle leaves into a temporary tent by weaving taut silken threads between them. Within this protective canopy they munched and grew, shedding their skins in small black piles four times as they matured. After about four to six weeks chrysalids began to develop as the caterpillars drooped tail first from a twig; there, within a pale-golden casing adorned with spikes of burnished copper, silver and black, their bodies began to transform and darken, emerging after about a fortnight.

One day Elizabeth and her supportive husband Michael heard a strange drumming sound on their lounge wall. By this time they had discovered it was safer to shift the chrysalids inside in containers to protect them from wasps and spiders. Caterpillars about to pupate would crawl away from the nettle plant and stick onto the walls or hide behind books. Their elegant lounge adorned with Elizabeth's landscape paintings had become a butterfly carnival. But that

CLOCKWISE FROM TOP LEFT: Common grass blue butterfly *Zizina otis labradus*; copper butterfly female *Lycaena salustius*; yellow admiral butterfly *Vanessa itea*; yellow admiral caterpillar.

odd drumbeat was a warning. A green vegetable bug had entered through the ranch-slider door from their deck. It made a bug-line for a chrysalis on the wall. It sought to suck the juices from the pupating creature within its golden casing. The strange noise came from the amazing protective response of the seemingly helpless pupa: it was wriggling furiously, swinging to and fro like a punch ball, almost at right angles to its point of adhesion to the wall, to drive off the predator. On the clump outside Elizabeth saw chrysalids exhibit the same strategy to repel wasps but, sadly, without success.

When she heard of the Monarch Butterfly New Zealand Trust she modestly told Jacqui Knight, one of its principals, that she was successfully breeding and releasing these rare and precious admirals into the wild. 'When we moved here seven years ago I established a cottage garden full of flowers around the house to encourage insect life for the orchard and for the general health of the garden. Later I planted a clump of stinging nettles – for nettle tea ...'

To the butterfly protectors this was almost unbelievable. When Jacqui passed on the news to Jan and me we set out for butterflyville with our cameras. We met a riot of nectar-producing flowers and a friendly couple and a cosy bungalow perched on a hillside with long views to the coast. The lounge of this house has become the realm of the butterfly. They emerge from their rebirthing pods on the piano, on the curtains, behind the chairs and on the book stack and they stick on to the walls. Predatory spiders are kept at bay. No fly sprays can be used. You must take care when you move. Elizabeth is utterly devoted to her progeny. She never knows what any day will bring forth; emergence is all quite haphazard. I noticed that as a chrysalis catches the light it gleams exactly like gilt foil. How *does* nature know how to mimic gold?

Out in her garden, as vivid as a coral reef, our cameras zoomed in on exquisite, newly opened wings drying in the sun and fibrillating periodically. A yellow admiral landed on the strap of Michael's watch. Then, to get the best light angle, I found I could induce one to delicately and trustingly strut from a deck rail onto my finger. Jan held one at arm's length on her index finger and focused her camera on it with the other hand. Their 80-year-old neighbour told them that when he was a boy in Auckland, red admirals were quite common, but the yellows were a rarity.

After hovering around her garden, feasting on a brilliant array of flowering plants, these rare, ornately patterned creatures, 'flowers with wings', flutter off into the beyond, about 300 over the summer season. It is just as if Elizabeth was breeding endangered native birds, releasing them to the wild. She is showing us how to restore the balance of nature. The pollinators of our exquisite New Zealand Eden, so drastically threatened by exotic wasps, are being replenished. But Elizabeth never got to drink her nettle tea. 'I might destroy an egg,' she smiled gently.

GLOW-WORMS

I could hardly believe we had glow-worms on our land. I thought they just lived in caves.

Coming home unexpectedly late one evening I found myself stepping from a car down at the gate and a long way from our hilltop home. I had no torch and it was a pitch-black night. As I groped cautiously up the bumpy shingle drive through the bush I suddenly became aware I had guidance … On one side the bank cutting was aglow with tiny blue lights. I could easily determine where the steep drainage ditch was; an obstacle I most feared on my dark journey. Glow-worms were adhering to the undercut clay bank. It never gets wet, no matter how heavy the rainfall. It is damp and sheltered from sun and wind. For the glow-worm, it is just another habitat. Even some of the old native timber fence posts in damp places on our borders glow at night.

Glow-worms
Arachnocampa luminosa.
BRADY DOAK

In caves glow-worms festoon rock walls, crannies and ceilings like stars in the sky. It came as a surprise to learn they are not worms at all, but the larval stage of a fly or gnat called *Arachnocampa luminosa,* and they are unique to New Zealand and Australia.

The adult insect lays about 100 eggs in clutches on the cave wall. In up to three weeks, they hatch into translucent larvae a few millimetres in length. The larva grows to about 40 mm within a horizontal silken tube that is suspended from a canopy of threads adhering to the roof. It moves up and down this woven tunnel, monitoring and repairing its fishing lines: multiple dangling threads. There can be up to 50 of them, armed with sticky beads of mucus that ensnare prey. Excreted from its tail, the attractant is a glowing substance, a mix of two chemicals that react with a blue-green luminosity, like the glow-sticks humans manufacture as night beacons. The struggles of trapped insects alert the glow-worm and it hauls the prey up to its mouth. In the stillness of caves ensnaring threads may reach half a metre, but those beneath clay banks such as on our land are only about two cm long or they would entangle. If disturbed the glow-worm can swiftly extinguish its light.

After about nine months the larva creates a cocoon in which, over a period of 12 days to three weeks, it transforms into a twin-winged adult rather like a mosquito, but slightly bigger at 15 mm. This adult never feeds. It has three days of love life and then it dies. If unfortunate, the parent could be caught and eaten by its own offspring, which just glows to show ...

LONG-TAILED BATS

When I heard of a bat colony just 10 km up the road from our gate, I was riveted. Bats are New Zealand's only native land mammals. They are threatened and increasingly rare. We have two species: the long-tailed and short-tailed bats. Those at Glenbervie are the long-tailed species or pekapeka (*Chalinolobus tuberculatus*).

The long-tailed bat is an insect eater and feeds on the wing after dusk. The rarer, slightly larger short-tailed bat is a night feeder on weta and other forest floor life as well as fruit and nectar. In this respect it is an important pollinator for certain natives such as the *Astelia banksii.* Agile on the ground, it is a poor flyer, seldom rising more than three metres. Both species use cavities in old growth trees for daytime roosting and shelter for their young. Loss of so much mature native forest puts their refuges at risk. Exotic pests also threaten the ground feeders. In the space of just seven days, one feral cat devoured 102 short-tailed bats around their roosting tree.

The long-tailed bat uses a form of natural sonar: it beams out high-frequency calls (beyond the realm of human hearing) and listens to the echoes in order

to locate aerial prey, such as midges, moths, mosquitoes and beetles, and to navigate around obstacles in pitch darkness as it flies at speeds of up to 60 km/h. A membrane along the bat's tail assists it in catching insects on the wing.

The bat usually spends each day sheltering within small tree cavities, but does not roost regularly in the same tree, so females must fly their clinging young from one to another. They produce only one pup each year, which makes their survival in our modern world even more precarious.

In the nineteenth century both species were abundant all through the country, in colonies of a thousand or more. Nowadays few New Zealanders have ever seen one and their ecological role is scarcely fulfilled. Jan and I feel hopeful that with increasing pest control and the maturing of native forests, our very special native bats may re-establish on the mainland. In this respect, our local bats could spread further afield. Recently, near a massive grove of old-growth taraire trees at Glenbervie, Jan and I watched for bats from sundown until the last squib of light. All the birds, including a pleasing number of native pigeons, had gradually gone to roost. Moths and beetles were fluttering and droning around us. We peered at the tree canopy silhouetted against the fading light. We knew bats fly with continuous flapping of wings with scalloped trailing edges, never gliding as do the swallows. We saw no bats at Glenbervie. The question remains; but ultrasound detectors prove they are there.

The short-tailed bat is the best pollinator for many natives such as the male *Astelia banksii*.

SHORE BIRDS AND SEABIRDS OF OUR COAST

NEW ZEALAND DOTTEREL / TUTURIWHATU

You could visit a beach frequently and never notice the New Zealand dotterel (*Charadrius obscurus*) – and the obscure bird would be perfectly happy to be ignored. Along the dynamic part of the shoreline where waves are withdrawing or among the drifts of seaweed flotsam a tiny brown ghost prospects for sandhoppers and other small creatures. It is a beachcomber. It cannot swim but runs like a flurry of wind and, then, it pauses stock-still. Jan and I only became aware of the New Zealand dotterel when it *wanted to be noticed*.

In mid-November we are strolling along a narrow path over the dunes at Woolleys Bay. Just ahead a bird is in distress: one wing is held out like a boomerang and it seems to have difficulty walking. A buzzer goes in my brain. This is the ploy the male dotterel uses at nesting time. Somewhere close by his partner will be huddled, incubating two or three brown speckled eggs in a scrape in the sand. As her protector he tries valiantly to distract any potential intruder with dramatic play-acting. After about four weeks their busy young will emerge. During their midwinter–spring nesting season the undersides of both genders flush rusty orange.

Visiting a fortnight later, close to the surf zone we notice fast, erratic movements of a downy ball of feathers; an animated cotton reel. Hither and yon three wisps of life are darting about the shoreline. Then they all converge on their mother to vanish beneath her warm feathers.

One of our zoomed pictures, enlarged on the computer, revealed a leg tag. I sent it to national dotterel expert John Dowding in Christchurch. To our delight he recognised the bird: 'It definitely came from Waipu, about a 49 km straight-line distance to the south, which is fairly typical – most don't return to their natal site to breed. It is seven or nearly eight years old. Average adult life expectancy is about 14 years, with a few birds lasting into their twenties.'

Once its presence is noted the New Zealand dotterel is quite easy to observe if you are in the right area. But this is an endangered species. Only about 1500 exist, because invaders threaten their nesting sites: people, dogs, rats, stoats, cats and subdivisions; whole suburbs of beach homes have been built on the dunes where dotterels nest.

During their nesting months the couples are widely distributed, each pair

within their own territory. But this changes once their young have been raised. In February we took pictures of a flock of 20 New Zealand dotterel roosting at high tide on the Te Maika sandbank near Ngunguru River entrance. The adjacent Ngunguru Sandspit/Pimanu narrowly escaped becoming a housing zone after a 40-year battle by the locals to save it from developers. From spring 2011, much of it came into public ownership; Pimanu could become a very special place for the dotterel.

When I reported this aggregation to Dowding he responded: 'Ngunguru Sandspit is a well-established post-breeding flock site for New Zealand dotterels. Numbers in these flocks usually peak in mid-February to late March, so you may count a few more yet.

'Based on criteria in the International Ramsar Convention, reflected in the recent New Zealand dotterel recovery plan, a site that regularly has 17 birds or more is considered of national and international significance. Under the current NZ Coastal Policy Statement, that, coupled with their Nationally Vulnerable status, means that actual or potential impacts on Ngunguru Sandspit New Zealand dotterel should be avoided.'

Our native birds that nest on the foredunes of our beaches face another

OPPOSITE, TOP TO BOTTOM: Male New Zealand dotterel *Charadrius obscurus*; a dotterel performs a fake broken wing distress display; courtship coloration (both genders develop this); protecting chick on the beach.

ABOVE: On Te Maika sandbank near Ngunguru River entrance, New Zealand dotterels and godwits roost at high tide.

threat: where exotic dune plants have been introduced, such as marram grass, nesting success may be compromised. Where native plants such as pingao and spinifex survive, nesting is more successful. The exotic dune plants can't withstand salt-water inundation and die back, creating a steep-faced dune where the sea sweeps away nests. Our native plants thrive with salt-water exposure, resist the erosive forces of wind and sea and actively grow down the beach berm to create a gently sloping dune foot, ideal for the nests of New Zealand dotterel and oystercatchers. Plant and bird have evolved together.

OYSTERCATCHERS / TOREA

Feathered opera singers. *Huep, huep, huep,* skirling down Ngunguru Estuary at twilight, the shrill piping alarm cries of an oystercatcher. Along Matapouri Beach three large black waders trot along together. Then a territorial dispute erupts. *Kerveee, kerveee, kerveee,* they cry, swerving about, orange beaks to the sand like hockey players. There is to be no threesome …

With their bold colouring, powerful beaks and strong legs, pairs of variable oystercatchers (*Haematopus unicolor*) defend shoreline territories on estuaries, harbours and beaches. They forage at low tide for well-anchored worms and bivalves. At high tide they roost together for security at places such as the crests of sandbanks. About two-thirds are totally black with the rest having a variable pied pattern.

Couples form long-term pair bonds. From around the start of August they tend two to three green, black-mottled eggs in a sand scrape on a small ridge. In early February we watch a juvenile mewling to its mother for food at the water's edge. It has grown almost as big as her, but its legs and beak are paler.

In autumn and winter on the roosting site at Te Maika sand cay, just within the Ngunguru River entrance, we notice a separate group of oystercatchers. The stand-aparts are all pied and slightly smaller. Pied oystercatchers (*Haematopus finschi*) breed in the South Island in spring. In December they migrate to feeding grounds in the northern North Island.

CASPIAN TERN / TARANUI

With its black cap and bright red bill the Caspian tern (*Hydroprogne caspia*) is the largest of New Zealand's terns. From time to time we see a few roosting with other birds at high tide, or feeding up the Ngunguru Estuary. Like other terns they are fish eaters, diving on small surface-feeding species such as sprats, herring and piper. In mid-December on extensive sand dunes at Mangawhai Heads, 50 km to the south, we found a colony of 25 with a few grey-capped juveniles. Up to three eggs are incubated by both parents in a sand scrape nest.

Nest guarding nearby were fairy terns, white-fronted terns and southern black-backed gulls: a rich area!

WHITE-FRONTED TERN / TARA

Another visitor with a similar lifestyle is the white-fronted tern (*Sterna striata*), the most common tern in New Zealand. At Ngunguru we see small groups in summer. About an hour's drive away around our west coast harbours, such as the Kaipara and Hokianga, we see flocks of them. Feeding opportunities handy to their chicks may be better in harbours. On an impossibly windy day we watched a parent hunting prey for a pair of fledglings waiting with plaintive cries on a concrete slipway. Under very difficult conditions the parent managed to find sprats by flying very low over the intensely ruffled water. The hungry chicks were always fed in the same order. White-fronted terns are often called 'kahawai birds' because they feed on pilchards and small baitfish chased to the surface by schools of kahawai.

WHITE-FACED HERON / KOTUKU

Since 1941, when a group flew in from Australia, white-faced herons (*Egretta novaehollandiae*) have had huge success throughout New Zealand. They have such a wide range of habitat: we see them on the seashore, in estuaries, up rivers, on farmland. They feed on invertebrates, frogs and small fish

OPPOSITE, TOP TO BOTTOM: Variable oystercatchers *Haematopus unicolor* mating; nest-guarding on a sand crest; young bird pleading for food from its parent; South Island pied oystercatchers *Haematopus finschi* migrate north for food.

TOP TO BOTTOM: Caspian terns *Hydroprogne caspia*; a juvenile white-fronted tern *Sterna striata* receiving a sprat; white-faced heron *Egretta novaehollandiae* with nuptial neck plumage.

wherever they can be found, often stirring the bottom with one leg to disturb prey. Or stalking meditatively in the shallows surrounded by their own elegant reflection. I have noticed that individuals will partner with other species in a feeding symbiosis: the heron walks along a mangrove bank while a pied shag swims in parallel, each scaring prey to the other. I have seen it feeding with a red-billed gull, a pied stilt, a royal spoonbill and even a rare white heron.

In spring, when nuptial plumes develop on their necks, they build twiggy nests in coastal trees. One February a friend took a photo of grey downy chicks on a nest in an old pohutukawa close to Tutukaka Marina. Others nest high in forest trees on our river margins. Both parents incubate three to five blue eggs and guard their young.

REEF HERON / MATUKU MOANA

In bays of Tutukaka Harbour a solitary reef heron (*Egretta sacra sacra*) stalks prey along the water's edge, where tiny sprats cruise and crabs crawl. Such a dark blue it almost seems black, it crouches low just like a cat on the prowl. Reef herons are loners. At Ngunguru we see red-billed gulls attacking like dive-bombers a reef heron roosting on a rock. This heron is regarded as a threat because it enters nesting colonies to steal chicks. On the Tutukaka Coast in spring, on a ledge below a major black-backed gull colony, we zoom in on the untidy nest of a reef heron. Only one parent is attendant. About three young are raised.

PIED STILT / POAKA

Yet another species that has flown here from Australia, the pied stilt (*Himantopus himantopus*) has the grace and elegance of a living sculpture. It stalks daintily and silently in the shallows or across open paddocks, questing worms or other invertebrates with its rapier beak. I have filmed one in Ngunguru Estuary vibrating its legs rapidly to disturb prey in sandy mud.

In early spring mature pied stilts may leave our estuaries and head for their breeding colonies along South Island rivers and elsewhere on our coast, where they nest on mounds in protective colonies. While either bird may incubate three to five greenish eggs, the other is on guard and if threatened, performs the most elaborate play-acting to simulate wing injury, flopping about and crying with pain. They will screech and dive-bomb any approaching interloper. Like so many Australian migrants, they are well adapted to mammalian predators, and have been hugely successful in New Zealand, whereas our native black stilt of South Island rivers, lacking these protective adaptations, is critically endangered.

CLOCKWISE FROM TOP LEFT: Blue reef heron matuku moana *Egretta sacra sacra*; blue reef heron nesting on coastal islet, late August; pied stilt flock in flight; a pied stilt *Himantopus himantopus leucocephalus* stalks prey.

EASTERN BAR-TAILED GODWIT / KUAKA

In mid-September long, slender beaks cleave the air above the Pacific, homing on estuaries in New Zealand, guided by subtle navigational cues. The godwits (*Limosa lapponica baueri*) are returning after seven months overwintering and breeding in Alaska. On grassy tundra hillsides in May–June both parents have shared incubation of up to four eggs.

Until felled by quakes, a Christchurch cathedral bell would stand ready to peal for 30 minutes to celebrate the springtime return of these wading birds: kuaka – the eastern bar-tailed godwits. In 2007 female godwit 'E7' touched down bearing a tiny radio transmitter. News flashed around the world: a record-breaking, non-stop flight of 11,500 km in about eight days.

Keeping an eye on shrinking godwit populations has taken on new importance with the problems of climate change. Their feeding grounds in the tundra of Alaska and Siberia are threatened by global warming. For humans, such global birds are the canaries in the coalmine: their survival is an indicator of our own prospects on a small, crowded planet. At our end of their flight godwits face serious threats from disturbance by humans and dogs.

Godwit E7 had been radio-tagged at Miranda on the Firth of Thames, a major wading bird habitat. Concern about the spread of avian flu by migratory birds spurred this expensive study. Scientists say that a godwit on a long flight consumes eight times its basic metabolic rate or BMR: the energy used while resting. No human athlete can reach more than five times his or her BMR energy

Bar-tailed godwits kuaka *Limosa lapponica baueri* in flight.

output, and then only for a few hours. E7 holds the world record for long-distance, muscle-powered flight; endurance champion of the animal world.

Prior to their early March migration godwits build up body weight to over 55 per cent fat as an energy store. Then, to compensate, their digestive system shrinks by about a quarter so that when fuelled up they are not carrying excess weight.

One day when reviewing our Ngunguru godwit pictures, Jan and I spotted an orange leg tag with no numbers. I wrote to Adrian Riegen, national godwit expert, who informed us that it had been tagged in Victoria, Australia, while young.

A few weeks before the godwits' autumn departure Jan and I went to Ruakaka to meet them en masse. With the king tide they would all abandon their Whangarei Harbour roosts as the tide covered the banks. Ruakaka is their next best hope of a roosting site. On cue, three flights of godwits dotted the sky above the Marsden Point Refinery and whizzed towards us, banking briefly and touching down on the glistening strand at the river entrance. Through our zoom lenses the myriad spots on the strand resolved into a varied bunch with red-breasted, breeding-ready males, speckled females with their extra long, upcurved beaks and pale, slim juveniles. Many birds stood around in small shallow lakes, some with heads buried in feathers; one leg withdrawn. A growing chorus of bird voices swelled like a theatre crowd at interval; godwit wings increasingly raised and stretched overhead. I admired those surprisingly large wings, soon to beat continuously, day after day, on the long migratory flight to winter feeding grounds in Alaska. Then, with no warning, the whole bird crowd was airborne.

Godwits roosting at Ruakaka.

They wheeled overhead in a swirling storm of silhouetted wings, banked steeply and were gone, up the coast to Whangarei and their mudflat haunts, newly exposed by the falling tide.

PIED SHAG / KARUHIRUHI

A sleek arrow speeds eastward along the cliff face, downwind. It banks, turns and flies back lower down. Now the light westerly acts as an air brake. Approaching the colony's pohutukawa tree it swoops up so that gravity further reduces its air speed. Precise calculation enables the big diving bird to alight with webbed feet on a branch beside its nest. It is much like planes landing on an aircraft carrier; or me walking upstairs wearing diving fins!

The male pied shag (*Phalacrocorax varius varius*) has a twig to present to his mate – but there is a ritual greeting to perform as tactful as at a Japanese tea party. When he alights she opens her beak to him. She bows her head low like a swan. He gives her the stick. His pink throat pouch is expanded and his beak gapes wide. Light shines through his throat tissues. There is a blue ring around his eye and a sulphur-yellow patch above the base of his distinctive hooked beak. She begins weaving his offering into their large coarse nest. Then his wings spread wide and he swoops down towards the sea to gain speed. In about five minutes he returns with another stick. During his absence she is busy weaving. Nearby a more advanced nesting couple have young birds in their nest. There are many nests in the colony.

Pied shags *Phalacrocorax varius varius* building a nest.

These are diving birds; most of their diet is fish. In mangrove creeks and estuaries and at sea they hunt individually: making long dives, often surfacing

with a preyfish in the beak, quickly gulping it down before diving again. Out at the Poor Knights Islands, 18 m down, all the fishes may suddenly take cover. It's not a shark; nor a kingfish. Nothing seems to scare the fish community more than a pied shag, looking like a drowned turkey, streaking up the channel. Shags have been caught in nets at 60 m so they can really dive deep.

The pied shags nest here all year round with peak activity in August–September and March–April. Each couple will produce from three to five chicks. Eggs take about a month to incubate. Once Jan and I made a low-tide journey from Woolleys Bay to reach the rock platform below the colony. It turned out to be a risky mission, aiming our lenses directly up at the nests. Droppings from the colony whiten the rocks below. Unfortunately in some situations wastes from the colony may kill the host tree.

LITTLE BLACK SHAG / KAWAU-TUI

By the estuary entrance to Mill Creek, a scrap of tree held a trio of small shags. One was asleep, head beneath a wing. A fourth shag skidded in and joined them, holding its wings out to dry. I lay prone on the footpath to steady my camera, not having my tripod handy. Jan took a gamble and crept closer to them, draping

Little black shags *Phalacrocorax sulcirostris* roost on a dead branch.

herself over a pohutukawa stump. Our cameras clicked as beaks swivelled and occasionally made pleasing symmetries. When we got home our pictures, downloaded to computer, confirmed a discovery: zooming in we found we had a quartet of little black shags (*Phalacrocorax sulcirostris*), originally an Australian species and a new sighting for us on Ngunguru Estuary. I liked the marbled pattern on their furled wings. Of all our New Zealand shags they really are the swarthiest: even their beaks and feet are black. They often feed cooperatively, much like dolphins, herding prey upriver in line abreast underwater. They breed in spring, usually in trees above fresh water.

ROYAL SPOONBILL / KOTUKU-NGUTUPAPA

The royal spoonbill (*Platalea regia*), related to the ibis, is one of the most exotic wading birds to self-introduce to New Zealand. At Onerahi Peninsula near Whangarei Airport, their flock has expanded dramatically in recent years. The last time Jan and I visited them we counted some 50 birds. And now a trio has taken up residence just down the Ngunguru River from our home: on the beach near Cape Horn and the old cemetery.

On their elongated legs spoonbills patrol the shallows in small groups, sweeping their long, flat, spoon-tipped beaks in arcs through the water. Sensors on the beak interior enable them to catch small prey: small fish, crustaceans and molluscs. At high tide they roost together for protection on a sandbank.

From September to December at nesting sites they build a platform nest of sticks lined with plant material. Following an elaborate courtship ritual involving raising and lowering of special breeding feathers on the head, beak clapping and exaggerated bowing gestures, the female incubates two or three off-white, brown-blotched eggs for 21 days. Both parents feed the hatchlings for around seven weeks. In autumn, flying strongly in a V formation, necks extended, they disperse to feeding areas around harbours, estuaries and lakes. Trees in Rangaunu Harbour may have up to 265 spoonbills roosting in their canopy at high tide.

The first spoonbill I ever saw on the Ngunguru River, back in the 1980s, was feeding symbiotically with a white-faced heron that accompanied it along the shore, snatching up prey items disturbed by the wading spoonbill.

ABOVE: Royal spoonbill kotuku-ngutupapa *Platelea regia* pair grooming on a tree crest.

OPPOSITE, TOP TO BOTTOM: Royal spoonbills in flight; a red-billed gull *Larus novaehollandiae* colony on an islet.

RED-BILLED GULL / TARAPUNGA

In spring red-billed gulls (*Larus novaehollandiae*) form a colony cresting the islet at Woolleys Bay. Nests are closely packed and there are disputes for a site near the centre. On the skyline we see wings flutter boldly as a male mounts his gull friend to continue the species. Up to five eggs are laid and both parents share the

task of incubating and raising fledglings. Before exotic predators arrived, gulls could have nested anywhere along this coast.

Gulls feed very widely over the open sea, along the shores and out in paddocks, but in estuaries the red-bill has a droll strategy. It jitterbugs, stamping its red webbed feet in muddy water at a frantic pace, moving ever so slowly backwards to expose hidden prey. My microphone picks up the *pitta, pitta, pitta* sound, like splashing raindrops in a squall. Red-ringed eyes search the mud. Occasionally the dance routine disturbs a tiny organism. The beak dips in a flash.

SOUTHERN BLACK-BACKED GULL / KARORO

During spring Jan and I discover a nesting colony of big southern black-backed seagulls (*Larus dominicanus dominicanus*) on an isolated rock close inshore near the North Gable.

The black-backed gull colony is on a golden rock. Their high-rise apartment is resplendent in a carpet of orange lichen. To our amazement, in the deeply indented cliff below the gulls, beneath an overhang, on a sunny ledge, is the untidy nest of a reef heron. These birds are quite rare; to find a nest is even rarer. But to see one right beside a colony of gulls on whose chicks it preys – that is mind-boggling! Ironically these gulls prey on small birds too.

A puzzle for the shore explorer is ngoiro, the juvenile black-backed gull. The large brown-and-white bird may exceed its parents in size. Until its third year this scavenger hardly looks like the same species.

GREY TERNLET

Near the Poor Knights Islands grey ternlets roost on ledges that riddle the sunny north-west cliff wall of the Sugarloaf. They flutter out briefly and settle back in their refuges as our vessel *Perfect Day* hovers nearby on a calm autumn afternoon. These delicate little seabirds remind me of doves. This is the only colony I know of in the area and they are quite rare in New Zealand.

The grey ternlet or grey noddy (*Procelsterna cerulea albivitta)* is blue–grey, with a paler body that becomes progressively darker along the wings. Its short tail is deeply forked. There is a small black patch in front of its eye, and the slender black bill is slightly curved. It has a soft, purring call. There is no visual difference between males and females, or in breeding plumage. Juveniles have a brownish upper body and crown. Nests are in hollows and pockets on cliff faces and consist of a few shreds of seaweed or grass.

Grey ternlets are found in the tropical and subtropical Pacific from mainland Australia, Lord Howe Island, Norfolk Island and east to Hawaii and certain islands off the coast of Chile. About 17,000 pairs breed on the Kermadec Islands. Some years they also bred on the Three Kings and other islands east of the northern North Island: Maori Rocks at the Mokohinaus and Volkner Rocks near White Island.

The term 'noddy' may refer to their courtship exchanges but also meant simpleton. When Darwin first met them on the famed *Beagle* cruise of 1831–36 he wrote: 'The noddies, as their name expresses, are silly little creatures.' With all due respect to the great scientist, I just hope our species lasts as long as the noddies! Grey ternlets are a threatened species in Australia and endangered by intensive fishing pressures. Their choice of the Sugarloaf for a colony protected by its marine reserve status was not too silly. In the days when I owned a fast dive boat we could reach the Sugarloaf over calm water in 12 minutes from Tutukaka entrance. So I count grey ternlets as my neighbours.

OPPOSITE, TOP TO BOTTOM: Red-billed gull *Larus novaehollandiae*; colony with red-billed gulls mating; black-backed gull *Larus dominicanus dominicanus*; grey ternlet *Procelsterna cerulea albivitta* in a colony on Sugarloaf Island, Poor Knights Marine Reserve.

GANNET / TAKAPU

In 1973 scientist Dr Walter Starck climbed to the summit of the Sugarloaf, about 100 metres above the ocean. He wore only rubber sneakers, shorts and a T-shirt. In a kitbag on his back he carried a 16 mm movie camera and a still camera. While I helped crew his ship RV *El Torito* around the huge perpendicular rock, he filmed us framed in the neck of a juvenile gannet. Around us was a vast school of baitfish and trevally. The guano-whitened island is home to the biggest gannet colony in this part of the north. A small side-shoot is housed on the adjacent Pinnacles. None live on the forest-clad Poor Knights Islands. Australasian gannets (*Morus serrator*) need a treeless, barren rock stack devoid of encumbering trees for their nesting colonies or ganneteries.

BELOW, LEFT TO RIGHT: gannet *Morus serrator* in flight against the sky; a gannet searches above foaming water for prey.

In most bays along the eastern coast of Northland, its harbours and estuaries, a gannet on broad wings spanning 180 cm can be seen circling 30 metres up, in quest of prey. Suddenly the wings fold as it plunges at up to 100 km/h into the sea to bob up with a sprat in its beak. Occasionally a cloud of gannets will work aggregated baitfish in Tutukaka Harbour, Ngunguru Estuary or the entrance to Ruakaka River.

In springtime common dolphins round up meatballs of baitfish. Cyclones of gannets circle and dive, panicking the anchovy herd so that dolphins can pick off fleeing prey while maintaining the dense sphere of fish: a complex symbiotic feeding pattern further complicated when up to three species of whale may join in, herding the fish against their vast bodies as they swim on one side, jaws agape, common dolphins on their bows and birds plunging like dive-bombers. Packs of bronze whaler sharks sometimes circle the meatball much as do the dolphins. At peak feeding intensity the gannets may stay below, alongside the meatball, beaks probing into it after prey.

In spring, mid-September to mid-December, gannets lay a large single pale blue egg in a nest of seaweed, which takes about 44 days to incubate. Fed frequently by both parents with nutritious regurgitated fish, the chick grows so fast that within about 14 weeks it exceeds its caregivers in size! In February–March, after days of exercising their wings, the plump, mottled young birds migrate to Australia, an initial uninterrupted Tasman flight of about 2000 km. Some may range as far as Fremantle in Western Australia. Like our wandering young folk, for the next two to six years they feed in Australian waters before returning to their birthplace, finding a mate and bonding for life, which may last up to 38 years.

ABOVE: A parent gannet feeding regurgitated fish to its young.

OPPOSITE, LEFT TO RIGHT: A gannet begins its dive for preyfish; mid-descent; about to enter; splash!

Blue penguin *Eudyptula minor* on its nest, Tiritiri Matangi Island sanctuary.

BLUE PENGUIN / KORORA

At dusk by the wall stones protecting the inner harbour at Tutukaka a small, ocean-blue, flightless bird emerges from the sea and scuttles into a cavity among the rocks. Six weeks earlier, she laid two speckled white eggs within. Now she regurgitates food for her month-old chicks. Her mate will follow and assist. They live in a permanent pair bond and return to their nest site each breeding season. (Otherwise they stay at sea for months.) At first light both birds swim far offshore, as far as the Poor Knight Islands (25 km), to dive 20 metres or more for small baitfish, shrimps and squid.

Sometimes night fishers on the end of the rock mole shudder to hear humanoid cries from within the wall. Blue or little penguins (*Eudyptula minor*) can sound just like a baby wailing.

One morning, as our dive boat left port, we saw a penguin fleeing over the surface, chased by a small shark. Neither could close nor widen the gap. We could not stay there long enough to learn the outcome.

At Tiritiri Matangi Island sanctuary in mid-October we were able to get otherwise impossible pictures of a parent guarding eggs in a specially made, window-topped nesting box. Blue penguins need help where there are exotic predators such as cats, dogs and stoats. These are the smallest penguins in the world.

AUSTRALIAN PELICAN

In spring 2012 amazing news greeted our ears: pelicans flew to Northland from Australia. (Eight months later, in autumn I received a picture of 11 pelicans on the shore of Rangaunu Harbour. All New Zealanders now. Will they breed in spring?)

Some years back I am at Altona Bay in the upper reaches of Port Phillip Bay, Melbourne. Groups of red-billed gulls are busy feeding among the eel grass. Each bird is tugging at a fat, juicy sand worm. Some are on bare mud, others beneath a few centimetres of water. The tide is falling. Up cruises a big Australian pelican (*Pelecanus conspicillatus*) like a sailing ship. The gull sets the worm aside and moves away. Time after time I see the same behaviour: gull extracts prey, pelican paddles in and eats it. Every so often a gull gets to gulp down a worm while the pelicans are preoccupied. There is no dispute; no conflict. Might is right. The pelicans are using the ability of the gull's beak to pull worms from the mud. Their own huge beaks would be too clumsy for such a task. There are about 50 pelicans in this part of the bay. Nature has a live and let live formula.

On 6 August, 14 pelicans arrived on the Northern Wairoa River and established themselves near Ruawai. The following morning one was sighted in an inlet near Kerikeri, perched high on a mooring pile. This may be the same bird that later visited Ruakaka, Waipu and Mangawhai.

These are huge birds. The males are bigger, with up to 2.5 m wingspan and 13 kg weight; the bill, at up to 50 cm long, is the world's largest. At take-off they kick both feet on the water. In flapping flight they don't travel more than about a kilometre. But on warm days, seeking new feeding areas, they gain height by spiralling on thermal updrafts, just like gliders. They may reach an altitude of 3000 metres where they form a V formation, perhaps 100 birds, to glide afar on the wind at speeds up to 56 km/h. Over the years since 1890 a few individuals have flown the 2000 km over the Tasman to New Zealand; once a trio came; but none survived long term. One, mistaken for a swan, was shot. Once birds reach here under their own power they qualify as natives.

Breeding can take place at almost any time of year where food supply is good but is mostly in late spring in cooler areas. Personally I can't wait to see the flamboyant courtship display. A female walks in a circle. A group of males follow in a courtship parade, swinging their huge bills from side to side and tossing sticks and objects into the air, catching them again. Both sexes clap their bills and rapidly ripple their pouches, which at courtship time become bright salmon-pink to advertise sexual readiness. Eventually the most impressive male is selected by the female and they set about nesting in a breeding colony, usually protected by surrounding water. One to three eggs are laid in a simple nest – a scrape lined with seaweed, grass or feathers – and both parents incubate them for about a month, on or under their feet. After an initial feeding period the

On a pile in Kerikeri Inlet, Australian pelican 'Picky' *Pelecanus conspicillatus*.

young are cared for in a communal crèche.

Hunting is communal, too. Several birds will encircle a small school of mullet or herd them into the shallows, scooping them up. Their bill pouches can hold 13 litres.

On Friday, 14 September 2012, Jan and I watch with rapture an Australian pelican feeding and resting on the west shore of the Northern Wairoa River. It opens its massive beak and shakes its pendant pouch with vigour. Behind, on the mangrove-fringed bank, stands a sleek white heron. A short distance upstream another 13 pelicans are feeding in groups on grey mullet.

Fears that the Tasman travellers would starve on the Kaipara Harbour have been dispelled. When we reached the riverbank three Tongan men were just carrying away a fish crate brimming with fat mullet. They gave us a couple of whoppers. Dargaville kayaker Pat Miller reports seeing mullet leap and accidentally striking several with his paddle, and he saw one disappear into a pelican from the opaque river. Provided mullet can thrive and their sensitive breeding grounds in salt marshes are protected from wandering cows, we may see these pelicans multiply and spread to estuaries on both coasts. They may become our neighbours at Ngunguru.

Nearby we filmed a flock of over 50 snowy-white cattle egrets. They are a sky poem as they wheel and billow above the fields. Breeding in Australia, they regularly commute across the Tasman.

Kevin Wigmore has just given us his pictures of the solo pelican that visited Ruakaka. I believe it is the one that went to Mangawhai. It's probably the same one that called briefly at Waipu Cove and may well be 'Picky', the pile-roosting pelican that arrived at Kerikeri on 7 August. Did it overfly our Ngunguru home? Pelicans may become a future inhabitant of our district and a familiar part of the New Zealand landscape.

The massive beak of an Australian pelican.
TONY AYLING

CLOCKWISE FROM TOP LEFT: A pelican visits the Ruakaka estuary; pelican takes off. KEVIN WIGMORE; cattle egrets *Bubulcus ibis* on the levee at Ruawai.

FOREST & WATER BIRDS OF THE TUTUKAKA COAST

KIWI RETURN TO OUR LAND

At 4:40 one calm July morning an exchange of kiwi calls took place near our hilltop home. A male called loudly and at length, an upscale sequence of about 20 calls; a female a little further away responded with a shorter, more guttural cry. There was a pause of five minutes and then this sequence was repeated, seemingly a little further away. Shortly after, a shrill, poignant male call.

Of late and with growing excitement at many locations on our land and adjoining properties we have been documenting kiwi beak probing holes. Some are up to 13 cm deep; often angled and showing the beak shape. I was interested to learn females have longer beaks than males: 13 cm versus 10 cm. This would give them access to more food items. The kiwi has a powerful sense of smell, unusual among birds, and is the only bird to have nostrils at its beak tip.

Clearly, this being their mating season, a kiwi courtship is in progress. Perhaps an offspring will result. I hated hearing a car on the highway; two roadkill kiwi have been found on the road verge near here in recent months. One was a mature female. Her genetic value: inestimable. Backed by DOC, we are getting road signage installed.

The previous Thursday, 23 June, at 6:30 pm, Jan first heard a kiwi call on our land. She heard another at very close range in the Trounson Park and we have recordings of them supplied by Landcare Trust. We have often heard them in the nearby Crawford Reserve over the years.

There is now no doubt that at least one kiwi has crossed the highway onto our land from the Crawford Reserve. With the Riverlands Landcare pest-controlled area of some 70 hectares, bounded on two sides by rivers, and the highway to the north, kiwi would have a very large area to exploit. We have an auto camera set up to get pictures. Our biggest worry is wandering dogs. A dog trap has been set.

As I lie in bed thinking of that North Island brown kiwi (*Apteryx mantelli*), probing our hillsides with its powerful beak, it comes to me: these are such ancient creatures, strands of the Gondwana tapestry of life, trying to survive in a human-dominated world where so much has vanished: moa, huia, giant eagles: some 57 New Zealand bird species have become extinct since human

OPPOSITE: Mike Camm cradles the brown kiwi *Apteryx mantelli* in his arms.

KIWI RELEASE AT MIKE AND JANE CAMM'S PLACE

Ecologist Dr Ray Pierce estimates there may be around 200 kiwi in the protected area around Mike and Jane Camm's place in the forested valley at the end of Waiotoi Road.

Waiting there in the bosom of the forested hills and valleys on Mike's land, on a tender, fecund spring day, warblers and miromiro singing, was a pleasure for the group of nature lovers, young and old, knowing a DOC vehicle was en route with three kiwi to be released in the vicinity.

Out on the coast was the headland at the entrance to Matapouri Bay and one arm of the Matapouri River has its source at the lake in the foreground – an artificial lake Mike and Jane have created, which spills over into a raupo swamp encrusted with coal-black freshwater mussels, much beloved by the local pukeko.

Eventually a grunty ute bumped along the farm track, driven by DOC worker Miriam Ritchie, a Matapouri local. On the tray in special wooden boxes were two female kiwi and one male. For Mike and Jane and their neighbours in the Tutukaka Landcare Coalition, this would be the culmination of their long efforts to rid the land of pests and to control wandering dogs. The small lake glinting in the afternoon sun is habitat for endangered pateke brown teal, which DOC has entrusted to Mike's group. Now, with an estimated 200 kiwi in the district, they are being given excess birds from an offshore island to the south, fulfilling DOC's plan to create breeding populations and release them back into mainland areas once protection is in place.

For the Year One and Two kids of the Ngunguru School this was a day they would never forget. Mike cradled a kiwi in his arms just like a baby, its head snuggled in the crook of his elbow, his hands clasping its powerful three-clawed legs. Each child got to stroke the coarse brown feathers. Miriam explained where its ear is, just behind the beady eye. Then she probed among its feathers to reveal the vestigial wing. The male kiwi had been fitted with a tiny radio transmitter because it is the males that establish the nest and guard the large solitary egg until it hatches. A fine example for New Zealand men, a lady said. Then many of the adults present became tender and parental, cradling a kiwi as it was admired and gently petted. 'Don't touch the sensitive beak,' they were advised. Eventually it was time for release. Mike stooped down with one and put its head into a prepared burrow in some fern. A directional aerial held aloft by a DOC worker traced the male's progress reassuringly.

Kiwi release day, North Island brown kiwi.

occupation. Kiwi are pretty tough birds in their own habitat and were once highly successful breeders, but in today's world of forest clearance and fast vehicles, dogs and exotic pests, six per cent of them are vanishing each year.

Like gannets, kiwi form permanent couples and can live for about 40 years. The female, with her egg demands, is larger: 40 cm tall. She weighs one-fifth more than the male. Because of their nocturnal lifestyle and steady food supply kiwi can breed regardless of the season, almost all year round: from June to March. Mating and male versus male territorial calls begin an hour after dusk and duets continue at intervals until an hour before dawn.

If the pioneering male on our land can attract a female from the Crawford Reserve to venture across the road she may have to find her way around the raupo swamp to reach the sanctuary he has found. But kiwi have been observed swimming! One had a territory intersected by a river, which it swam regularly. Thanks to pest control our land is now teeming with unexploited food resources. The kiwi diet is astonishingly omnivorous: stomach contents have yielded worms, snails, slugs, centipedes, millipedes, woodlice, weta, crickets, cicada nymphs, moth grubs, caterpillars, ants, glow-worms, spiders, beetle larvae, native cockroaches, koura or freshwater crayfish, sandhoppers, young eels and frogs; also seeds and berries such as those of the nikau, miro, kahikatea, maire, rimu, hinau, toro, and the many coprosmas and the seeds of hebe shrubs and

Gahnia grass. For almost all these food items modern kiwi have to compete with rats, possums and hedgehogs.

On our hillsides of regenerating native forest, he will follow her around constantly while grunting. It may be two months before she is ready to mate. When she is receptive he taps her on the back near her neck and she crouches low, her head resting on the ground. Unlike most birds the kiwi has a penis. He may grasp her feathers in his beak for balance. They may mate three times in a night. But if she's not interested, she might wander off or, being bigger than him, she may even try to scare him away. In captivity a spurned kiwi male has been seen to leap around with hopping strides, roll on his back and kick his feet in the air and then do somersaults downhill!

There are multiple sleeping burrows in a territory, but a new one is used for breeding. Well concealed and measuring up to three metres long, it has already been excavated and the end chamber, big enough for two birds side by side, is lined with plant fibres. Three to four weeks after mating she lays her egg in it. In relation to female body size the kiwi egg is the biggest in the world. Weighing up to 430 grams, it has taken up so much of her insides that lately she has been

People gathered for the release of three kiwi at Mike Camm's sanctuary for kiwi.

unable to eat. Now her duty is done she can resume feeding. But about three weeks later she may lay a second egg; more work for the male!

The father is especially fitted for egg incubation. At this time he has developed a brood patch of bare skin among his nether feathers to keep the egg warm. At intervals he turns the egg to regulate an even temperature. Each night he leaves the nest for a period of up to five hours to forage.

It takes about three days for the chick to break out of the egg. Its eyes are open and it is fully feathered. For a couple of days it gets nutrients from the yolk sac via its navel. Initially this navel linkage prevents it from standing. About a week after hatching the young kiwi's foraging life begins. Accompanied initially by its father, it is soon alone and very vulnerable. Unlike its nocturnal parents, it feeds by day and by night. After four to six weeks it leaves its parent's territory and within a year attains adult size and begins to call. Males call up to three times more often than females, especially early in the breeding season.

In the standard, pest-riddled habitat of today their chick has only a five per cent chance of surviving its first six months. After four months it is stoat-proof. A female may breed in her third to fifth year. Over a period of some 30 years or more she could produce about 100 eggs.

With five species scattered through wilderness New Zealand, from Northland to Rakiura/Stewart Island, we are fortunate to have the kiwi as our national icon. With such a very odd and special bird it is imperative from the point of view of the world that it should survive. At Christmas a camper on our land stumbled upon a kiwi in the dark. Both were equally startled and uttered quiet *oohs*.

NEW ZEALAND PIGEON / KERERU

On an afternoon in early October I am down our sunny, sheltered drive where I have been working for weeks to improve it with pick, spade and shovel. My reward is a cacophony of birds: New Zealand pigeons (*Hemiphaga novaeseelandiae*) blundering about in crazy courtship; shining cuckoos giving those loud, insistent mating calls; tui fluttering about fussily; and a male miromiro singing so sweetly. Next day, mid-afternoon, Jan and I take our cameras for a walk. Twice, at different locations, we meet a pigeon pair; in each case, sharing the same trees.

For at least two months, in December–January, the native pigeons dine several times a day on the fruit of our female matipo trees. These are forest-margin, dioecious trees: the matipo males outnumber the food-bearing females greatly. Tui and pigeons know where every berry producer is and they fight over them. Jan films a tui and a pigeon in combat right by our house in a fruit-laden matipo tree. The local tui drives all birds away from *his* territory. He keeps zooming overhead

LEFT TO RIGHT: Courting New Zealand pigeons *Hemiphaga novaeseelandiae*; a New Zealand pigeon on a branch digesting berries.

with glancing contact. Just above the perching fruit pecker it elaborately displays widespread wings, alternately perching or hovering, but no go. The plucky pigeon will not budge. Eventually they flutter off, pigeon in pursuit.

In autumn, an hour before dusk, a flock of 12 pigeons circles above our lawn seat on the forest hilltop. Why such an aggregation? I notice several flights of tui heading towards a tall grove of trees. That's it! I follow their direction. The female kahikatea tree is a tall cone of golden fruit. For days the birds gorge on this sweet bounty. Once Maori men laboured to harvest these small berries.

With the onset of winter I see the pigeons grazing the tender leaves of lacebarks. In spring they eat the new leaves of our kowhai.

As the largest of our seed-distributing birds, pigeons – kereru or kukupa – are essential to New Zealand rainforests. So many of our major trees produce large berries that only they can manage. On our forest land regeneration is spectacular because pests no longer eat these berries, leaving an abundance for the birds, whose diet includes fruit of puriri, tawa, miro, matai and others. Around certain tall trees we see a circle of young karaka, taraire and pigeonwood. The seedlings even manage to overcome long grass, so persistently do the pigeons release seed as they perch on a branch digesting their food. In forests where these birds are now absent scientists find regeneration has fallen by up to 84 per cent. Mature trees eventually die off naturally. Without replacements the New Zealand forests are collapsing.

TUI

One autumn Jan and I were visiting our neighbour Pat Fitzgibbon's place on a ridge adjoining the Crawford Reserve. Beside his house he has an extensive orchard, with a range of pip and stone fruit. I could not understand how the trees could survive depredation by rosellas, blackbirds and other fruit-stealing birds. To our amazement we learnt that the secret was a tui (*Prosthemadera novaeseelandiae*) who would not allow any other feathered competitor into the area. From observations around our own home tui are the dominant birds. For orchardists and vintners tui could be special allies provided they were allowed some fruit. They are very territorial and quite widely spaced, but their survival depends on our eliminating rats and stoats and other bird predators. With intensive local pest control Jan and I have noticed a huge resurgence in tui and a mosaic of territories has developed.

The tui belongs to the honeyeater family of birds, along with the bellbird, and is well equipped for feeding on the nectar of flowers. But this is an adaptable bird; nectar is not always available and we see it feeding on a wide range of food sources, from insects to berries and exotic fruit. Jan has even seen one hop to the ground to prey on a snail just like a thrush. During a drought we found a juvenile

OPPOSITE, TOP TO BOTTOM: A juvenile tui, with orange beak and white collar, sips nectar in a pohutukawa bloom; an adult tui *Prosthemadera novaeseelandiae* at a water trough.

drowning in a partial bucket of water. It could not escape. The weakened bird responded to honey offered on our fingers. A very long thin tongue was thrust well beyond the curving black beak; a snake-like brush-tipped tongue well adapted to probing the tube of a flax flower.

In spring tui build nests in the densest parts of our totara trees: large structures of twigs and straw and far beyond camera access. They lay two to four eggs, white or pale pink, blotched reddish brown at the thick end. The male is the larger and his twin white throat tufts are more prominent. The female has a paler russet abdomen.

Like pigeons tui have a major ecological role in distributing the seed of smaller berry trees and pollinating flowers. As their numbers rise with pest control we note increased fecundity in the forest. Pests take their food and destroy their young.

ABOVE: Bellbird *Anthornis melanura*.

OPPOSITE, TOP TO BOTTOM: A fantail *Rhipidura fuliginosa* watching for disturbed insects; A fantail singing; A fantail clings to a trunk, seeking insect on wing.

BELLBIRD / KORIMAKO

Springtime in our district is abuzz with bellbird stories! James Johnston saw one near his mother's home opposite the Ngunguru Hall. Scattered reports of them come in from along the coast, especially in Mike Camm's pest-controlled valley.

Extremely abundant on the nearby Poor Knights Islands, bellbirds (*Anthornis melanura*) rarely nest on this coast. Their feeding and nesting habits are similar to those of the tui, but their diet has more emphasis on insects. We dream pest control may restore their sweet chimes to our dawn chorus.

FANTAIL / PIWAKAWAKA

Can fantails count? In spring I watch a parent feed three young a large insect it has smashed to pulp on a branch, as they pester for food with mewling sounds. It is passed to them to peck off a portion, one chick after the other, always in the same order. Then off through the forest the adult flits, followed by hungry progeny. Feeding of fledglings is mostly by males.

Fantails (*Rhipidura fuliginosa*) build a small saucer-shaped nest. Both parents brood the creamy speckled eggs and care for up to four chicks, raising up to five clutches over a season.

The friendliest of birds, these agile little insect eaters share our lives so intimately. They follow our walks, seeking insects we disturb. They have entered our buildings; one even landed on our seated son's knee. But these are among the hardest birds to capture well with a camera, so seldom do they rest for more than a second. We enjoy their aerobatic flights, especially against the dawn or dusk sky; their cascades of bright sound; the broad spread white tail-fan, with its two central black feathers, that acts as an air brake as they flit with the utmost agility from twig to twig. When cicadas are dinning in our ears they will even attack these large prey and, standing on one leg, consume it in the grasp of the other. Whiskers flanking the beak help to intercept prey. I have seen a fantail snatch an insect on the wing with its claw. While we sit on our deck an individual will come to our generous birdbath only a metre away and leap in repeatedly to cleanse its nether regions. Fantails are said to live only three years.

SILVEREYE / TAUHOU

In 1856 flocks of silvereyes (*Zosterops lateralis*) began migrating to New Zealand from southeastern Australia. Having arrived here under their own power they now qualify as natives. Many of our birds got here that way – and most of our fishes!

I notice that they fly around in flocks during autumn and winter, but in spring they pair off. Males have brighter colours than their partners. They swarm around our blossom trees sipping nectar. With some flowers they peck an access hole in the base, spoiling the plant's chance of pollination. Some natives, such as the lovely orange gloxinia, are becoming scarce because of this. The trumpet blooms need the attentions of nectar specialists like bellbirds; sometimes a tui will do.

On our deck we maintain a broad, saucer-shaped birdbath made by our grandkids. Whereas most birds visit it individually, silvereyes bathe in groups of four to six. And while some birds, like fantails, wash only their rumps, silvereyes duck right under repeatedly, turn by turn.

Attracted by our neighbour's breadcrumb handout these birds crowd around Pat's porch, vigorously quivering their wings while close to each other. This is an aggressive behaviour.

Silvereyes perform great service in devouring small insects and grubs on our trees. They distribute widely the small berries of native trees and shrubs. They peck holes in our figs, passionfruit and persimmons as soon as the least ripeness appears.

In spring I am clipping the *Abelia* hedge that forms a tunnel over our drive. My blades reveal a small, elaborate nest with three pale sky-blue eggs. Close by the distraught parent silvereye flutters angrily. Up to three families can be raised in a season. Both parents feed the chicks, bringing them insects, fruit and spiders.

When our feijoa trees, a South American relative of the pohutukawa, were in flower I noted a silvereye parent nipping off those pale-green, plump

oval calices from which the crimson feijoa blooms emerge. Three chicks were beside it on the branch. One by one they were fed the flower cases. I tried one: pleasantly sweet. It would be a valuable baby food.

GREY WARBLER / RIRORIRO

Shrill piping sounds gave them away. Somewhere along a path through the bush to our house a baby bird was seeking parental attention. Then a delicious sequence of mellifluous notes, candy for the ear, rang out. A grey warbler (*Gerygone igata*) fluttered from branch to twig like a hummingbird. In mid-air it hovered beside foliage, a ruby eye hunting for grubs. In a deft series of aerobatics and beak clackings it snapped up a small moth on the wing then down it swooped to its offspring wobbling on a fallen stick at ground level. I rushed inside for my camera.

Pretty soon I realised there were two warbler chicks begging for food beside my path while their parents flitted frantically about, as busy as shop hands at rush hour, trying to satisfy tiny appetites that knew no end. One chick fell

OPPOSITE: On our birdbath rim, silvereye *Zosterops lateralis,* dives in . . . emerges, dishevelled but cleaner.

ABOVE: A silvereye feeds on coprosma berries.

LEFT TO RIGHT: Grey warbler *Gerygone igata* parent at nest entrance; chicks in nest.

off a log, clinging upside down by its claw like a baby monkey. After a while I assessed this was the clumsy one of the pair. For some reason these baby birds had abandoned their cocoon-like nest before they learnt to fly. Their parents had nested early enough to avoid the parasitising of their nest by the recently migrated shining cuckoo, which would very soon lay an egg with the next warbler broods and that chick, hatching first, would toss out the warbler eggs. On this early set of nestlings and a late-season brood, the grey warblers depend for survival.

Jan came home from the Ngunguru Medical Centre where she works. Sandy Page, the physiotherapist, told her they had a pair of grey warblers build their nest close to their house. When a cuckoo hatched it perched on a twig beside the nest. The foster parent warblers found they could not alight on the twig to feed their adopted offspring, so they would hover in mid-air to deliver food to its beak. Of all the birds in New Zealand the grey warbler is unique in being able to hover like a hummingbird. This requires high calories and a small body.

A warbler's nest is right on the ridgeline path to our house, clearly visible in a manuka bush with sheltering *Abelia* hedge all around it: an extended oval structure with a hole in the north side just above halfway. This would be the best-constructed and most weatherproof nest of any bird species we have seen.

Just then a tiny grey warbler flew into view, hopped on a series of twigs and then zoomed into the hole to reappear in a second and fly off. For Jan and me this was the chance we yearned for: to monitor a warbler's nest. Not in all the years since we settled on our land in 1978 had such a chance arisen. While warblers are abundant here their nests are usually very hard to see unless disused and dislodged by a storm. At Whale Bay, Jan had almost fallen off the cliff trying to focus her zoom lens on a nest.

We set up our cameras. This was a feathered husband and wife team. Like carpenters they had initially created the basic structure; now they were doing the lining. Each time they returned from a building materials foray, they usually had a small white downy feather in the beak or perhaps a scrap of spider web. One bird would enter the interior while the other patched the outside cladding. We were surprised to see the couple usually worked in unison. This would mean that if one had found useful nesting material it would wait to synchronise their joint return. This may be necessary to ensure a coherent effort. I have read that nest building takes about a week.

The day before I'd noticed a patch of small white feathers by a neighbour's puriri tree; scene of a bird attack. Now all those feathers had vanished and our neighbour Pat De Silva told me she had seen birds gathering hairs from around

LEFT TO RIGHT: A grey warbler chick being fed by its parent; a grey warbler chick awaits food.

Lady, her moulting Alsatian dog. (I said to her dignified ladyship, 'Did you know you are becoming a bird's nest?')

Jan put tassels of sheep's wool near the nest and watched the warblers busily snatch bits up and weave them into their nursery. To her surprise there was not the least hesitation in accepting her offering. It may assist them greatly as they spend up to 40 minutes before returning with a small feather. Perhaps it is a race to get a clutch away before the cuckoos arrive. When they ceased to use the wool we assumed construction was complete and nesting must have begun.

After a stormy night we noted that the nest had an elaborate and well-made addition. There was a porch nicely contoured into the overall form to shelter the entry hole. But following a major blow we were very worried. The nest site is on a ridge where the wind is especially violent. I had even considered it as a site for a wind generator. The nest survived, but the porch had been closed down like a shutter. We glowed with delight when we saw the nest shuddering. Now, we thought, eggs may be developing beneath small grey wings.

Jan and I were mindful that up in the tropical Bismarck Archipelago, north of New Guinea, some slightly bigger birds with emerald feathers and a black-and-white striped breast would be hunting insects to build up energy reserves for their long flight back to New Zealand in a few weeks' time.

Each year Jan has kept in her diary a record of when the distinctive calls of shining cuckoos signal their return to the trees near our home. The migrant female cuckoo will lay an egg in a warbler's nest. Incubation for this egg is around 12 days (for grey warblers, about 18). The baby cuckoo expels its rivals. Then the warblers feed it diligently with insects and grubs for another 12 days until it is fledged and for a further four weeks thereafter, as the chick follows the foster parents around pleading incessantly. But our little warbling songbirds can produce several clutches of eggs over the summer. They reproduce themselves very successfully in our area and with their help the new cuckoos are able to mature for their migration to the tropics in February.

Watching it all you become aware of the interconnectedness of life on this planet. The native forest that shelters both bird species is all the healthier for the effective pest control those busy beaks exert. And our tiny, drab New Zealand bird with the most plaintive and melodious of songs plays a role in the health of distant tropical jungles. The little insect eaters are one of the most abundant birds on our land. Of all the local birds, they build the most protective of nests.

SHINING CUCKOO / PIPIWHARAUROA

I nearly jumped out of my bird hide in the bushes. Overhead was a grey warbler's nest. Distinctive cocoon or nursery pod, dangling from a mingimingi twig. I had just decided the nest was abandoned. The three tiny chicks Jan and I had watched grow, day by day: all had gone. A sudden flutter of wings. A shining cuckoo landed at the back of the nest. She peered in through the lattice of straw and feathers. Prying eyes and sharp beak. She moved around the outside, but she did not approach the opening. No need. She had assessed that this was an empty nest. Off she flew in quest of another nest to parasitise.

Clever, I reflected. Had she tried to enter the nest through the portal there could have been a conflict with an egg-minding warbler parent. The cuckoo would have no advantage in fighting with the species she must rely on to hatch and raise her young.

Each year in mid-September Jan and I listen for the return of the shining cuckoos (*Chrysococcyx lucidus*). From the Bismarck Archipelago these small emerald bundles of feathers navigate to our shores. Over the summer months they mate and lay eggs in warbler nests, before departing in February for their winter feeding grounds.

Nobody has ever seen a cuckoo's nest. For a migrant bird, this is a clever strategy. No need to spend effort gathering nest materials. No time wasted hunting for food to satisfy the appetite of a chick. But successful parasites must not harm their hosts. The warblers still manage to flourish.

All summer we enjoy the distinctive calls of the cuckoos in the ridgeline

TOP TO BOTTOM: A shining cuckoo's banded feathers; shining cuckoo *Chrysococcyx lucidus* at rest after a long journey from the tropics.

TOP TO BOTTOM: Female North island tomtit *Petroica macrocephala*; male tomtit.

trees on our land where they hunt for leaf-munching grubs. On fine days they get together in a rock band. Their multiple calls merge into a cacophony of excitement; a noisy bird 'parliament'.

NORTH ISLAND TOMTIT / MIROMIRO

Jan writes: 'This morning I went down to lay rat poison around our bottom lane. I thought I'd lug my camera and tripod along – just in case … Three days ago Wade had taken Karen, our daughter-in-law, for a walk down there. Wade had been showing Karen our rare *Fuchsia perscandens* when they had a visit by a miromiro or North Island tomtit. They came back animated and raving, saying, "It came so close – the dear little bird, just sat at eye level looking at us!" … I heard the miromiro calling. I tried to imitate his call. I stood by the fuchsia, quietly, thinking: I wonder if I will see him. I noticed a fine mahoe branch swaying up and down and thought to myself that it looked as if something had landed on it; maybe the miromiro is there somewhere? Suddenly he just appeared on a branch in front of me. I managed to take three photos. With two of them he had started to take off, but initially I got one good one with him sitting there, looking straight at me, his blue eye, his black head and his strongly contrasting white chest. Taking photos of birds is a bit like taking fish pix underwater. You only get one chance …'

Mirimiro (*Petroica macrocephala*) and fantails are both insect hunters. But whereas the fantail catches prey with aerobatic flight manoeuvres, aided by its special tail, the miromiro has a very different strategy. It clings to a branch or twig and searches through the forest. When an insect is seen miromiro swoops directly and catches it in mid-air to land at once on a handy perch. Because of this hunting pattern in Maori lore it was held that miromiro would deliver a message between lovers. I like the high value this puts on the small forest dweller.

When a miromiro pair started to come around our home and geckos clung to the insect screens on the ranch slider or took up residence in our outbuildings and I nearly stood on a large gleaming kauri snail by our door – we *knew* pest control had made a major difference.

WELCOME SWALLOW / WAROU

A pair of welcome swallows (*Hirundo neoxena*) has built its convoluted mud nest under the overhanging roof of our tool shed. Usually these little aerialist insect eaters choose a man-made structure, but Jan and I know caves along the coast and out at the Poor Knights Islands where they nest in the wild – usually places where predator access is difficult. We enjoy their swooping aerobatics as they harvest a wind eddy on our hilltop. They remind us of plankton-feeding demoiselles beneath the sea. Both have forked tails for nimble turns and twists, and eyes close to the mouth for accurate interception of prey.

One afternoon in early January, at a birthday gathering on the south bank of Ngunguru River, we were joined by a huge flock of swallows that swarmed overhead in a whirlwind for some time. I can find no mention of such activity in New Zealand records. Old World swallows form aggregations in autumn prior to migration. We had no idea that there were so many swallows in our locality. When mosquitoes and sandflies are on the wing, these little immigrants are most welcome. They arrived in Northland in the 1950s, possibly while migrating to or from Tasmania. In Australia their return is welcomed as a herald of spring, hence the name.

Up to five eggs are incubated by the female for around three weeks. Then both parents feed their young. Once they can fly we have seen them accompanying a parent to be fed on a perch, as do fantails.

NEW ZEALAND PIPIT / PIHOIHOI

It is so easy to overlook the pipit. Its light-brown camouflage plumage and close resemblance to the European skylark both conspire to make me ignorant of this native bird as part of our local community. But the pipit never takes off to hover in the sky and sing. Out on open grass its presence among female sparrows and chaffinches is betrayed by frequent flickings of its tail. And above its eye there is a distinctive white brow. The New Zealand pipit (*Anthus novaeseelandiae*) is not as shy as the skylark and it does not feed on seeds as do those meadow birds it most resembles. It is an insect eater.

I mostly notice pipits when out on a country road. Something is hopping about among the dusty stones inspecting each pebble with unusual interest. Or

TOP TO BOTTOM: Welcome swallow *Hirundo neoxena* offspring (left) with parent; New Zealand pipit pihihoihoi *Anthus novaeseelandiae* seeking insects.

it may be silently questing sandhoppers along the driftline of a beach.

Like the lark it is a ground nester, which in today's pest-ridden world can be very risky. From August to February it breeds in long grass or bracken, laying several clutches of up to four cream, brown-blotched eggs. Both parents tend the young for about 10 days. Pipit were once much more abundant; early settlers recorded them roosting on their house roofs.

KAKA

On offshore islands such as Hauturu/Little Barrier thrives the kaka (*Nestor meridionalis*). It is New Zealand's third true parrot, beside the kakapo and kea. At Whangarei Heads, where intensive pest control is ongoing, a colony has established of perhaps a dozen birds or more. Stray birds visit the Tutukaka Coast, but they need security from predators and old-growth cavity trees for their nests. Ironically, Tutukaka means 'parrot perch' . . .

Having met flocks of them at island sanctuaries Jan and I long for the day when they can safely return to our locality. Such colourful and charismatic birds would be a fitting reward for years of effort by our community. We see how they stand one-legged on a branch to eat food with the other claw; we delight at the brilliant colours beneath their wings; their excited harsh calls and whistles. Kaka have huge versatility, able to reach into flowers for nectar with their brush tongues or use their strong beaks to tear open tough kauri cones for seeds, or logs for huhu grubs.

In winter they breed. Both parents join their efforts at raising up to four young. The vulnerable 90-day incubation time, when the mother is very much at the mercy of stoats and possums, results in a gender imbalance. Meanwhile lone males wander the coast, waiting for the time when it will be safe to nest here. Kaka are birds of our future.

NEW ZEALAND KINGFISHER / KOTARE

Our kingfisher (*Todirhamphus sanctus vagans*) has a very specific style of hunting. It just sits in wait for prey. Whether on a lofty power wire above our lawn, a mangrove branch or a small rock beside the estuary it maintains its low-energy vigil. It is scanning for the least movement by a lizard or worm, fish or crab. With lightning flight and rapier beak it snatches quarry and resumes its perch just as quickly, mostly without landing.

Kingfisher populations need nesting hole opportunities, such as clay banks or the trunks of old trees. Good burrows are reused year after year, September to January. In spring I have seen a trio engaged in a noisy courtship ritual among the branches of our gum trees. The female incubates four or five white eggs for

TOP: Feeding on figs, kaka, our native parrot *Nestor meridionalis*.

BOTTOM: A kingfisher *Todirhamphus sanctus vagans* seeks prey at river's edge; a kingfisher on a branch at the forest margin.

around 24 days and the male helps supply food to the chicks.

Where nest sites and food are abundant, kingfisher populations are doing well. They can defend themselves and their fledglings against most predators. On hot summer days the regular shrill cries of the kingfisher proclaiming its territory are part of our ambient music.

MOREPORK / RURU

In a shady glade down in our Magic Valley one summer noon I meet the morepork or ruru (*Ninox novaeseelandiae*). In the dimness the owl is just a silent wafting shadow flitting ahead of me. It flies as soundlessly as a thought. On a dark branch it roosts and faces me. It is most unusual to see this nocturnal predator by day. I think I must have stumbled on its dormitory. Once I sought shelter from sudden rain beneath the boughs of a huge puriri in our valley. Just as I arrived a ruru alighted on a branch close to my head. As its eyes adjusted to the gloom the pupils dilated. The intake of that gaze and those huge golden orbs held me spellbound in another time. Perhaps the owl knew this was a rain shelter within its territory – or its nest site.

The ruru is incredibly adapted for night hunting: special soft-edged feathers for silent flight; a head that can swivel 270 degrees; a face like a sound dish; enhanced ear channels; binocular eyes and a long, aerobatic tail. With stealth its sharp talons pounce on insects, small birds, rodents and lizards, mostly at dawn and dusk.

BELOW: Morepork *Ninox novaeseelandiae*.

OPPOSITE: Australasian harrier hawk *Circus approximans*.
NICOLA SHARMA

In spring the female nests in a tree hole, raising up to three chicks assisted by her partner. The first English sailors camped on Ngunguru shores, seeking kauri for ship masts, thought the strange owl was calling *haul taut*.

HARRIER / KAHU

Another bird of prey, the Australasian or swamp harrier (*Circus approximans*), colloquially called a hawk, is classed as a New Zealand native; it has been here for around 1000 years.

One summer day, just 10 m from our house, a hawk swooped down right in front of Jan and snatched a ground-foraging blackbird. It had trouble taking off with its prey because there were confining trees so close about. In a vertical lift it took three attempts. We think the hawks are feeding on baby quail and pukeko and the nestlings of other natives whose populations have been boosted by our pest control. They get fewer roadkill possums for the same reason.

I breathe a sigh of relief for all our local nestlings when I see a hawk circling higher and higher in the sky. Just like a glider it is gaining height to soar away downwind, a low-energy way to travel across country. And then I reflect: as a predator it plays an ecological role. If our local birds overpopulate and exceed their food supply they may nest in places where the hawk can readily find their chicks or eggs.

Hawks begin courtship displays from June with elaborate sky dances, circling and gliding. From September to December females build platform nests in raupo swamps where they breed and share the raising of their three to five chicks with the male, who delivers food to his consort.

SPUR-WINGED PLOVER

Wings widespread, the striking yellow-wattled male stands his ground and threatens me as I walk over the grassy hillside. The sentry's black-tipped golden spurs are erect and menacing. I back off. Each spring the spur-winged couple take turns to guard and incubate a clutch of four khaki-brown, dark-blotched eggs on this regularly mown sward. Reggie only discovers their rudimentary nest, just a hollow in the grass, after his ride-on mower has passed over. But this never deters the persistent parents. Now I realise why they are so successful in driving away hawks and other predators. Ground breeding in the open is very risky, but the noisy Australians inherit just what our endemic birds lack in defensive adaptations. Since they flew to our shores some 80 years ago these aggressive wading birds have flourished despite all the cats, rats, mustelids and mowers, adapting their shoreline lifestyle as invertebrate predators of the abundant worms and insects of New Zealand's extensive pastures – as have other immigrant wading birds such as pied stilts and white-faced herons.

Spur-winged plovers (*Vanellus miles novaehollandiae*) give joint care to their young and may bond for life – up to 12 years – and they live within a home range of about four km. In autumn and winter it is said they join flocks of up to 100, but I have yet to see this around here. At this time they may gather in small groups on a hillcrest to perform unusually complex display rituals, strutting in a circle or parading to and fro in a row.

FERNBIRD / MATATA

A claw clutching a twig; a beady eye and a tiny sharp beak; a fluttering dark shape in a dense thicket fringing the marsh. Once, on the crest of a swamp ribbonwood bush, after hours of waiting with my video camera, I recorded at extreme zoom range a small brown bird twittering sweetly. No bird, no species of wildlife under the sea or on land has so frustrated our efforts to obtain a visual record of its existence. The fleeting and delicate beauty of matata (*Bowdleria punctata*); the subtle tones and textures of its streaked and spotted chestnut-brown plumage spur us on; it *must* be possible, one day, to get our own pictures of this elusive native bird. We can hear it readily enough, as we creep along the pest control track we have made skirting a swampy stream on the margin of our land. We even play its call through portable speakers in the hope one will expose itself to our lenses for a moment.

To keep in touch with its lifetime mate and maintain its territory the bird emits a sound at intervals. It is a low-frequency *U* tone followed by a sharp metallic tick like a Geiger counter. During their spring breeding season the male initiates and the female responds with a click in a brief little duet.

One thing seemed certain: in this land of ground birds we would never see a fernbird flitting through the sky. Nipping about in quest of insects in the low, tangled marsh growth it seldom takes wing for more than about 50 metres, 100 maximum. Its distinctive long tail, half the 18 cm total body length, droops unaerodynamically like a wisp of straw. But for the matata this feeding style works very well: it is adept at finding moths and spiders and flies, grasshoppers and weta, bugs and beetles: all the myriad invertebrate life of the marshlands.

Starting in springtime (late August) a matata couple will construct a tidy cup-shaped nest, well concealed at the base of some rushes. About three tiny pale-pink, dark-speckled eggs are laid and both birds take turns to incubate them for two weeks and a similar time feeding their chicks.

For all our frustration, I was pleased. The insect-eating fernbird has an excellent niche on the margin between our swamp and forest. In a time when no mammalian predators stalked this bird world it was distributed throughout New Zealand. With introduced predators and habitat loss from land development its numbers have dwindled; it is now extinct in Canterbury. Exotic paper wasps nesting among the rushes compete with parent birds for food resources while they are busy raising their chicks. But predator control offers these special little ground birds a future.

Fast-forward to 2012. We wait in the marsh-edge forest for so long blowflies are checking us out speculatively. Something like a rat scurries through the reeds at my feet. A dark feathery shape tunnels out of sight. 'A fernbird,' I whisper hoarsely. Then the action begins. On every twig and fern frond, trunk, branch and bush our quarry flits on tiny wings. The long trailing tail like delicate fern tendrils; a beady eye; strong pink claws; complex patterns of dots and speckles and streaks. Suddenly there is a pair of fernbirds darting around us and overhead ...

OPPOSITE, TOP TO BOTTOM: Spur-winged plover *Vanellus miles novaehollandiae;* two spur-winged plover chicks on a hill-crest nest; fernbird *Bowdleria punctata.*

A banded rail *Gallirallus philippensis assimilis* bathes in mangroves.

BANDED RAIL / MOHO PERERU

I now realise just how lucky I was to get a photo of a banded rail (*Gallirallus philippensis assimilis*). Since then I have only seen them occasionally, dashing hell for leather across the road or scuttling into swamps like lightning. But when we lived in a small cottage so close to the mangroves we could see river life from our kitchen sink, we noticed ripples radiating across the water from the muddy bank at low tide. I decided to stalk whatever it was. One morning down through the mangrove trunks to the river edge trod a largish, spindle-shaped bird with a long tapered beak and broad clawed feet. We had never seen it in several years of living there. The stranger proceeded to wash itself in the brackish stream.

Banded rails tend to forage on mudflat margins at dawn and dusk, especially after a falling tide. Besides worms, spiders and insects, small fruits and seeds, a favourite prey is the mud crab, which emerges from its burrow at this time. During the middle of the day banded rails seek cover back in the saltmarsh reeds. And this is where territorial pairs breed from spring to summer. In a cup-shaped nest about five pinkish eggs are incubated by both parents, who care for their young hatchlings for up to two months before evicting them from the territory. A smaller relative, the spotless crake (*Porzana tabuensis*), is even rarer and more secretive. It has been sighted in our locality and is more common on the Poor Knights Islands.

AUSTRALASIAN BITTERN / MATUKU

Wetland birds often lead such covert lives we seldom see them. One summer's day I stepped into a dense reed bed fringing our estuarine valley. My heart leapt out of my body when a large brown bird erupted from the undergrowth at my feet and flapped away at a low angle. A bittern! I recalled seeing *Botaurus poiciloptilus* in a nearby marshy paddock in its classic pose: frozen still, beak erect, like a tall spindle, swaying in the wind. This is cryptic mimicry: among reeds it becomes invisible. It doubles as a form of defence and a method of ambushing prey. A determined observer once spent hours trying to get a sighting. When she eventually moved a bittern sprang out close by. Our friend Mike Camm has seen a pair of young bittern swimming on his artificial lake.

Along our swamp-edge paths we have learnt to recognise the low, kakapo-like courtship booming of the male during its long breeding season from June to February. Usually in November or December three to five pale-olive eggs are incubated on a low reed platform. The female raises her young alone.

Bittern prey on large insects, eels, frogs, rodents, koura and birds. They have even been seen baiting fish with small pieces of grass.

Observing bittern is made even harder by their endangered status (only about 1000 birds remain), their large territories (around 50 hectares per couple), their mottled camouflage and their dawn and dusk feeding tendency. Pest control and habitat protection are essential for the survival of matuku.

Australasian bittern *Botaurus poiciloptilus* at pond edge.
JOHN CRAIG

ABOVE: Pukeko *Porphyrio porphyrio melanotus* grazes our lawns for wild carrot.
BRADY DOAK

OPPOSITE, TOP TO BOTTOM: Pateke (female) *Anas chlorotis* night feeding on lawn; the pateke (male) with the radio transmitter;pateke group roosting in a mangrove creek by day.

PUKEKO / SWAMPHEN

First the rascals rooted up my neighbour's banana palms; the following year they demolished his entire feijoa crop and much of his citrus. Henry is not very fond of this native bird, recalling pukeko (*Porphyrio porphyrio melanotus*) originated from Australia about four centuries ago. When they took over the birdbath on our deck they polluted the area with feathers and droppings, a barbarism that tui, silvereye and fantail bathers never exhibit. Ever resourceful, Jan tried putting a special tub for pukeko over in the paddock. This worked! They always use their exclusive bathing facility and cause us no bother.

Despite predators, pukeko are very successful. Perhaps it is their group defence and their Australian mammalian adaptations. One summer Jan and I found a nest in long grass at the centre of a clearing. It contained a dozen light-brown, purple-speckled eggs. This would have be a communal nest, contributed to by several females. Their breeding season extends from August to March and incubation takes 25 days. Both genders have a collective approach to babysitting, too. Men who stumbled upon a group of chicks on a track were left with bleeding arms and heads. If people are too tolerant of these handsome, smart, strutting birds they may invade houses, attack cats and eat the eggs of ducks. Pukeko can fly, run, swim and – unlike ducks – perch in the crests of trees.

BROWN TEAL / PATEKE

Only in New Zealand! A forest duck; a duck that feeds in the dark. And a highly endangered endemic duck.

At twilight the pateke (*Anas chlorotis*) leaves its daytime communal refuge at the reedy edges of lakes, ponds and tidal creeks and sets out to forage in marshes and over damp paddocks, seeking snails and slugs; worms and insects; seeds and juicy morsels of clover – much as pukeko feed by day. In the ancient world its most serious predators would have been diurnal birds of prey like the falcon and the giant eagle. In today's New Zealand, bristling with night predator mammals from abroad, our very special duck, like the nocturnal kiwi and kakapo, is in trouble. Its defensive strategy has turned on it. Once among the most endangered waterfowl in the world, pateke are now on the increase due to intensive management. When this hit home to Jan and me we felt a huge responsibility for a bird whose space we share.

In June from a nearby pond came reports of pateke courtship: lots of noise and splashings. Once distributed throughout the country their populations have been reduced to three main concentrations: Coromandel, Great Barrier Island and, sporadically, from Ngunguru north towards the Bay of Islands. These east coast regions have in common a similar geology: steep forested hill country which often restricts farming and limits development. Habitat loss is their second major problem, beside fishing nets and cars. With fewer than 2000 pateke in these key areas, DOC has initiated a serious breed and release programme nationally to avoid extinction. But this can only be achieved in conjunction with intensive predator control. Their numbers crashed seriously some 30 years ago. Back in the 1920s around 500 pateke could be seen on one short stretch of the Mangere Stream that flows from the Pukenui Forest near Whangarei.

Over the months of summer and early autumn, pateke flock together for protection around ponds, pools, drains, creeks and lakes, preferring to have tree cover for safe roosting. During the months of winter and spring, within their strongly defended territory, a pateke pair can raise up to two clutches of four to eight eggs each season. The mother incubates the eggs for about four weeks. In his gaudy breeding plumage – iridescent green head and chestnut-red breast – the male drives off all invaders, including other ducks.

One late July morning, as I wrote this entry on the pateke, local conservationist Mike Camm rang. In 30 minutes Jan and I had to be ready with our cameras. He took us to sites on Waiotoi Road, Ngunguru and Clements Roads, Matapouri, where we watched and documented as he monitored his 18 recently released radio-tagged pateke.

Holding aloft a small antenna, he smiles. Reassuring electronic squeals show where each bird is roosting. If it has not moved after eight hours the squeal becomes a double shriek and Mike has a body to track down. We even got a photo of a new chick with parents. Another picture shows a female nesting at the base of a streamside tree; the male stands on guard alongside her. When they cruise up the full-tide Matapouri River towards us, the male takes the lead. In a short space of time and travel we watch pateke pairs on an inland creek and in a mangrove-fringed tidal estuary.

PARADISE SHELDUCK / PUTANGITANGI

At Kowharewa Bay in Tutukaka Harbour one fine day in early October a pair of native paradise ducks (*Tadorna variegata*) escorts its six fluffy grey and white ducklings down to the water's edge for their first swim. Then the offspring gather in a huddle on the warm stones while their parents bookend them, facing apart so they can survey all sections of the beach. Unusually, it is the white-headed female that has the brighter colouring than her swarthy partner. At Mosquito Bay near the Kaipara Harbour's South Head, before sunrise, we see a female chase a male down the beach. At Bowden Cove on Tutukaka Coast we watch a female fossicking for crustaceans on the pebbly beach. Out in coastal paddocks bordering valley floor streams, over the winter, we notice pairs of paradise ducks snipping at seed heads, clover and young grass shoots and prospecting for worms and insects. This is a coastal duck and a river duck. Even as chicks they can dive down and they fly with their wings underwater.

In August a valiant male bars our way on a morning quiet road, waddling along, wings widespread in front of our car. Nearby his partner is guarding their nest among rushes and sticks or perhaps up in a tree hole. For a month she incubates up to a dozen eggs while her lifetime mate guards the centre of their territory. A day after the ducklings emerge both parents escort them down to

the water. As terrestrial grazers, paradise ducks for the most part use water as a defence, even sending their young out into small waves while they defend them from the beach. For their first two months both parents zealously guard their fledglings. In their first year the youngsters feed in flocks. The following year males challenge each other and fight over females, before forming a permanent pair bond.

One day in early January, Jan and I come upon a large number of adult paradise ducks flocking on a coastal pond at Mimiwhangata Farm Park. In scattered male and female pairs, over 50 can be counted in one photo. We can't understand why so many mature birds are so close together. Later we learn: during their three-week moulting period these ducks cannot fly. They gather in this way on a body of water for defence. The pair of paradise ducks living on open pastures and lawns of our Riverlands community will be here for life. But they are not endangered and seasonal hunting is permitted. This reduces life expectancy to about two years. Otherwise they would live longer. While juveniles may range up to 200 km, adults are quite sedentary and usually spend their lives within about 30 km home range.

At Mangawhai we go to meet an ardent defender of the highly endangered

Paradise shelducks *Tadorna variegata* parenting in Kowharewa Bay, early October.

fairy tern: Jane Vaughan. She lives close to a mangrove estuary and wetland. Sitting on her deck beside her like a pet cat is a male paradise duck. He has adopted her since losing his mate. Several times Jane has tried to restore him to his kin, but he keeps flying back to her house.

But why, I wonder, are paradise ducks not endangered, as is the pateke? On reflection: these day-feeding ducks are not so much at the mercy of exotic predators; being shelducks they are one-fifth bigger; they have larger families; and a long breeding season; and with forest clearance, their pasture and grassland habitat has been extended. Being Australian in origin, they are adapted to mammalian predators.

From our extensive wilderness travels throughout Northland I now realise how especially important this Tutukaka Coast area really is for biodiversity protection, with so much varied habitat available to endangered species like pateke and kiwi. From 1968, we have lived here in homes at Matapouri, Tutukaka and Ngunguru. Over those years we have seen a significant change in land use from farming to lifestyle blocks. Along with an emergent culture of nature conservation and predator control has come an enormous amount of regrowth. Nature is so fecund here and old wounds are healing. A friend just showed me a set of idyllic photos from Stewart Island. This same recuperative process is under way right through New Zealand.

Jane Vaughan, bird lover, with a male paradise shelduck that adopted her home – uninvited.

OPPOSITE: Paradise ducks flocking to moult on a coastal pond at Mimiwhangata Farm Park.

Sandy Bay, Tutukaka Coast.

CONCLUSION:

TOWARDS GONDWANA – OR BIRD SILENCE?

For most of our evolution we humans have lived in a locality wherein we had detailed knowledge of, and names for, a huge range of life forms in the sea and on the land and even out to the moon and stars; arbiters of our food production, tool-making, healing, clothing, shelter, navigation and art forms. Uprooted from our past, we urbanised humans must be the first generations to lose touch with our environment, ceasing to recognise the life stories of the plants and the animals around us. For our future on a troubled planet, reintegration with the natural world that supports our lives may be essential to our survival. Understanding the systems we inhabit may enable us to develop a culture that fits into them rather than destroys our life base.

From the ancient Greeks earth scientists have borrowed the term *Gaia* as a new way of thinking about the life on our planet. I could also have used the Maori equivalent, *Papatuanuku*. For me *Gaia* represents our growing scientific awareness of the life process, all the interlocking components of our blue planet's living mantle; that co-evolving community of life forms with expanding levels of complexity from the first self-replicating living cells. In my recent memoir *Gaia Calls*, which was published in the United States, I explain aspects of this and claim that understanding *Gaia* is essential to our survival. This memoir extends that material.

THE SILENCE OF OUR FORESTS

Mount Manaia, summit.

Under the sea Jan and I fought for marine reserves and saw their immense success unfold through time. But now, as passionate divers for half a century, relatively new to terrestrial problems, we are haunted by the fragmented habitats and silent forests we have visited in Northland and all that is implicit in the ecology of this silence for the future of our country. *I get the feeling we are leaving New Zealand.* Perhaps some of us never really got here? We grow accustomed to daily experience in a realm of exotic plants and birds from beyond these shores that is of our own making. A friend told me of a recent trip to the South Island's magnificent Hollyford Track where he could look down a valley through the entire ecological sequence from alpine to sea coast – but it was bird silent.

We claim our native forests in international forums as carbon sinks. Unless they are pest controlled or predator-proof fenced, we endanger their reproductive systems: native birds, reptiles, bats, and insects. Therefore, they are not truly sustainable. The New Zealand rainforest and its fauna have a symbiotic relationship. They have co-evolved in a complex web of life wherein each is entirely dependent on the other. Just like the coral reef and its symbiotic algae and our bodies with their gut flora, without birds *et alia* our forests are *finite.* Only forests that are sustainable should have full carbon-sink value. In this light, state funding to predator-proof native forest remnants, such as Whangarei Heads, Mount Manaia and the Tangihua Ranges in my own area, is of national and global importance. An investment funded by fossil fuel taxes.

Maybe we New Zealanders should just mortgage ourselves to the future and get predator-proof fenced sanctuaries in place around our native forests wherever landscape makes that practicable such as coastal peninsulas; or empower volunteers to develop intensive pest control, as with the superb Ark in the Park in the Waitakere Ranges near Auckland, where several endangered species have been released. I believe the future will be very grateful. A series of protected wildernesses, stepping stones and corridors to tomorrow, would have a huge impact on the future of our country and its visitor trade: people bent on experiencing the South Pacific.

An old school friend of mine living in Germany, Michael Mence, has financed a whole series of protected 'reflection woods' with a project that issues gift certificates for birthdays, weddings, Christmas presents and as memorials to dead loved ones. New Zealanders could consider subsidising a version of such a project at national level.

A metre of predator-proof fence currently costs around $100. Intensive pest control is cheaper in many situations. At Trounson Park, a remnant of Northland west coast forest surrounded by farmland, intensive predator control has created a pest-free zone. Part of Puketi Forest has been so well controlled that endangered kokako, sweetest of songsters, were recently released there. A valley on the Tutukaka Coast is so well protected, kiwi are often released in it (see page 161).

This Tutukaka Coast that I have anatomised with such passion is repeated

all around New Zealand. Jan and I have explored all the major forest remnants in Northland, documenting them with sequenced photo essays and hoping in vain to find some refuge where birds sing galore. For inspiration we have visited many sanctuaries: the offshore islands Tiritiri Matangi, Little Barrier/Hauturu, Poor Knights; Matakohe/Limestone Island in Whangarei Harbour; Trounson Park and Aroha Island; protected areas such as Whangarei Heads and Puketi Forest. While there are local differences due to latitude and other factors, *all through New Zealand* many of the same plants and animals recur with the same vital linkages and face the same dire threats from exotic pests and habitat loss. DOC cannot cope with the slow decline of our biodiversity. It freely admits it is just reducing the rate of loss. We must have a collaborative approach with local bodies, institutions and individuals.

LOOKING FORWARD

Perhaps the challenge New Zealand's chief scientist Sir Paul Callaghan made, shortly before he died, may be accepted by the people who produced Sir Edmund Hillary and Sir Peter Blake and the great, nature-inspired art of the Maori. In his last public address Sir Paul left us with an inspiring challenge: can we eradicate pests throughout our island nation? A *New Zealand Listener* editorial of April 2012 concluded: 'The conservation battle is everyone's to wage. With leadership from entities like DOC, Forest & Bird and scientists, and input from as many volunteers as can be rallied to the cause, and as much private and public funding as can be raised, Callaghan's vision can be kept alive. To do so is not only our privilege but also our duty.'

Already this process is under way in a thousand areas of Aotearoa/New Zealand. Urban dwellers delight in taking their exercise along a chain of rat bait stations through nearby forests. Each day they are noticing enhancements as Gondwana returns. Few efforts could be more rewarding than ensuring our future. Otherwise, as Sir Paul predicted, we face centuries of combating rats and possums and stoats. I wish a cost–benefit study could be done. How much do pests cost our economy each year? With such unique treasures to protect and such a precious green mantle that has us envied by most of the world, we might set out to unite all the predator-proof zones we have so far created and extend this protection to our whole island nation. Every step we take along that path is worthwhile.

Dr James Russell, senior lecturer at Auckland University's schools of biological sciences and statistics, and recipient of the prime minister's 2012 MacDiarmid Emerging Scientist Prize, echoed Callaghan's words in March

TOP TO BOTTOM: Stitchbird *Notiomystis cincta* at Tiritiri Matangi Island sanctuary; Dr James Russell, conservation scientist of Auckland University, with a morepork chick.

2013 when he wrote about the prospect of ridding New Zealand completely of introduced predators. He put a notional price tag – about $24 billion – on such an operation, which might be rolled out over the whole country over a 10-year period. And on the subject of how it would actually happen, he had this to say:

> *On the main islands of New Zealand it is expected that predator-free status would not, at least not at first, come about from an orchestrated strategic plan, but instead through grass-roots community organisations already existing across New Zealand. These groups would create predator-free 'arks', and over time increase their halo of influence by expanding the scope of their programmes.*

As I finish this book I have just attended a meeting of conservationists to establish a predator-controlled Kiwi Coast Corridor extending from Whangarei Heads up to the Tutukaka Coast. Further down the track this Kiwi Coast Corridor project will thrust north to meet major efforts under way in the Bay of Islands, where all the lovely islands are becoming pest free. It will incorporate areas that already have a history of pest control and subsequent wildlife recovery. It will link up a mosaic of other efforts to the south made by Landcare groups and individuals, often supported by the regional council and the National Biodiversity Fund. Because of the abundance of wildlife habitat, and the nationally endangered status of so many of its very special life forms, it is hoped that overseas wildlife funders and business corporates will assist.

Ngunguru Estuary sunrise.

It gets better: the World Wildlife Fund has recently visited Northland and wants us to participate in a major landscape connectivity project called Reconnecting Northland along with our local Kiwi Coast Corridor linking kiwi protection groups from Whangarei to the Bay of Islands. (For a parallel story, visit http://www.gondwanalink.org/.) It's a step that could be extrapolated nationally towards the fulfilment of Sir Paul's dream. Perhaps we might call it the 'Callaghanisation of New Zealand'.

Jan and I have just made our third family visit to the predator-proofed sanctuary at Tawharanui on the sea coast an hour's drive north of Auckland. We were elated at spending a day amidst a mosaic of bellbirds, saddlebacks, native pigeons, whiteheads, banded rails, brown teal/pateke and other birds vanishing from our shores. As we drove away up the winding hill at day's end, the impact of it all really hit home. 'Ring dingaling, ringdingaling bellbird, calling me all the day long,' that poignant old song, came to mind, our ears still chiming with their melodies, as if we'd been to a music concert. Tawharanui sanctuary is an outstanding and evolving success of the first order. 'New Zealand, you ain't seen nothing yet!' I exclaimed, with general agreement.

Naia, our 10-year-old grand-daughter – who has planted many trees at Tawharanui, regularly helps her folks inspect its fence, and assisted at the first release of saddlebacks there, and has taken many amazing photos; and visited other sanctuaries such as Trounson Park, Aroha Island, Tiritiri Matangi and Rangitoto islands, Maungatautari in Waikato and Zealandia in Wellington, Goat Island and Poor Knights marine reserves, and our place, often – just sent us her artwork encapsulating it all. The future restorers of wild New Zealand are in our schools. They approach it with zest. We really think it can be done. We hope this book has shown what is at stake and what there is to gain.

Naia Doak's artwork.
NAIA DOAK

OPPOSITE: Beach picnic at Tawharanui, a sanctuary below water and above, in its forests and wetlands. Welcome to the real Aotearoa/New Zealand!

FURTHER READING

BOOKS BY THE AUTHOR

Early diving days: *The Elingamite and its Treasure* (1969), *The Burning of the Boyd* (1984), *Ocean Planet* (1989). **Marine invertebrates:** *Beneath New Zealand Seas* (1971), *The Cliff Dwellers: an undersea community* (1979). **Marine fishes:** *Fishes of the New Zealand Region* (1972), *Wade Doak's World of New Zealand Fishes* (1991), *I Am a Fish* (1999) *A Photographic Guide to Sea Fishes of New Zealand* (2003). **Dolphin/whale research:** *Dolphin Dolphin* (1981), *Encounters with Whales and Dolphins* (1988), *Swimming With Dolphins in New Zealand* (1994), *Friends in the Sea: solo dolphins in New Zealand and Australia* (1995). **South Pacific Islands marine life:** *Sharks and Other Ancestors* (1975), *Islands of Survival* (1976), *Deep Blue: a South Pacific odyssey* (1997). **General:** *Gaia Calls: South Sea voices, dolphins, sharks and rainforests* (2012).

OTHER AUTHORS

Intertidal zone

Bradstock, Mike, *Between the Tides: New Zealand shore and estuary life*. Auckland: Reed Methuen, 1985.

Morton, John, *Seashore Ecology of New Zealand and the Pacific*. Auckland: David Bateman, 2004.

Morton, John, and Michael Miller, *The New Zealand Seashore*. Auckland: Collins, 1968.

Plants

Dawson, John, and Rob Lucas, *New Zealand's Native Trees*. Nelson: Craig Potton Publishing, 2011.

Poole, Lindsay, and Nancy Adams, *Trees and Shrubs of New Zealand* (revised edition). Lincoln: Manaaki Whenua Press, 1994.

Salmon, J. T., *The Native Trees of New Zealand*. Auckland: A. W. Reed, 1980.

Terrestrial invertebrates

Early, John, *Know Your New Zealand Insects and Spiders*. Auckland: New Holland Publishers, 2009.

Birds

Davies, Detlef, *Where to Find Birds in Far North New Zealand: a site guide*. Self-published, 2011.

Falla, R. A., R. B. Sibson and E. G. Turbott, *The New Guide to the Birds of New Zealand*. Auckland: Collins, 1966.

Robertson, C. J. R. (ed.), *The Reader's Digest Complete Book of New Zealand Birds*. Reader's Digest: Surry Hills, NSW, 1985.

nzbirdsonline.org.nz

Map showing the Tutukaka Coast.

ACKNOWLEDGEMENTS

Avian experts: Tiffany Browne/DOC (pateke), Clea Gardiner (kiwi) and Adrian Riegen (godwits). **Botany:** Guy Bowden, Lisa Forester, Tony Foster and Dorothy Hansen. **Early diving mates:** Keith Gordon, Kelly Tarlton, Jaan Voot and Jim White. **Freshwater fishes:** David Bruno and Mike McGlynn. **Friends and neighbours who have helped:** Elizabeth Amber, Brian Angliss, Dr Tony Ayling, Tony Beauchamp, Phil Bendle, Mike Camm, Ross Clark and Janet Burgess, David Clarkson, Tim Damerell, Steph and Bob Davidson, Pat and Henry De Silva, Brady Doak and Dr Karen Tricklebank, Naia Doak, Sophie Edwards, Kent Erickson, Russell Fransham, Colin Gans, Jeff Gibbs, Pat Heffey, Reggie Hole and Neal Rastrick, Melanie McDonald, Kevin Mathews, Michael Mence, Steve and Fiona Moase, Kura Newall, Nina Pivac, Nicola Sharma, Dr Walter Starck, Mark Turner and Raewyn Rouse, Hilton and Melva Ward, Paratene Wellington, Michael Wiese, Kevin Wigmore, Chris Windust and Beverly Woods. **Geology:** Dr Jeremy Gibbs. **Hauturu/Little Barrier Island access:** Judy Hanbury. **Land invertebrates:** John Early/Auckland War Memorial Museum and Milen Marinov (damselflies). **Land restoration:** Steve Henderson and Ngaire Tyson. **Lizards:** Benjamin Barr.

INDEX

Numbers in bold type refer to images.

ABOUT THE AUTHORS

Born in Canterbury in 1940, Wade Doak has from an early age been passionate about exploring the natural world. He began below the waves, designing and constructing his own dive gear, and in 1969 discovered a stash of coins in a shipwreck, which became the subject of his book *The Elingamite and its Treasure*. Treasure salvage allowed him to devote his life to studying and photographing the 'blue planet', and he has published 19 books about the ocean and its inhabitants, including *A Photographic Guide to Sea Fishes of NZ* (New Holland). Most of these books are lavishly illustrated with Wade's own photographs. With his wife Jan and son Brady, Wade worked on the television natural history series *Wild South* and *Deep Blue*.

Jan Doak grew up on a Banks Peninsula sheep farm, and a family passion for beach picnics and collecting seafood meant that when Jan heard about the new sport of skin-diving, she was eager to learn more. Meeting Wade, she found a willing teacher and their shared enthusiasms led to marriage, a honeymoon diving in tropical New Caledonia, and making a home next to the clear waters of Northland. A fully trained nurse, Jan found the precise demands of nursing were readily adapted to those of extreme close-up camera work and she became an accomplished underwater photographer. Jan and Wade have two children, Brady and Karla, who are both divers, and Brady works as an underwater camera operator.

In 2012, Wade was awarded the Queen's Service Medal for his services to marine conservation. He has been a strong advocate of Marine Reserves and especially for his beloved Poor Knights Islands, off the Tutakaka Coast in Northland.

Wade and Jan Doak on a scuba diving trip in the late 1960s, at Mayor Island off the coast of the Bay of Plenty.

Heart of our rainforest:
reflecting pool.